THE EXISTENCE OF
SPACE AND TIME

THE EXISTENCE OF
SPACE AND TIME

by Ian Hinckfuss

CLARENDON PRESS · OXFORD
1975

Oxford University Press, Ely House, London W. 1

GLASGOW   NEW YORK   TORONTO   MELBOURNE   WELLINGTON
CAPE TOWN   IBADAN   NAIROBI   DAR ES SALAAM   LUSAKA   ADDIS ABABA
DELHI   BOMBAY   CALCUTTA   MADRAS   KARACHI   LAHORE   DACCA
KUALA LUMPUR   SINGAPORE   HONG KONG   TOKYO

ISBN 0 19 824519 X

© *Oxford University Press 1975*

*Printed in Great Britain by
Richard Clay (The Chaucer Press) Ltd.
Bungay, Suffolk*

To
Delilah, David,
Thomas, and Kim

Preface

THIS book is intended as an introduction to the philosophical problems of space and time, suitable for any reader who has an interest in the nature of the universe and who has a secondary-school knowledge of physics and mathematics. It is hoped that the book may, in particular, find a use in philosophy departments and physics departments within universities and other tertiary institutions. The attempt is always to introduce the problems from a twentieth-century point of view. That is to say, problems are introduced and considered because they are relevant to the way we think of space and time today and the way we are capable of thinking of space and time today, with all the hindsight now available to us. The approach therefore is heuristic rather than historical. The beginner in this area, as in so many areas, would, I believe, only be confused by a historical approach. For the conceptual problems facing Aristotle and Galileo are not always the same conceptual problems facing the man in the street today. Often, conceptual innovations introduced by the philosophers of the past are ingested today in infancy. They have become a part of the cultural heritage handed down from parent to offspring—they have become 'second nature'.

This does not mean, of course, that the history of the philosophy of space and time is not important, let alone irrelevant, to our problems today. The reverse is the case. The point is that, given the aim of tackling the problems that face us today, it is preferable to introduce the history of the topic if and when that history becomes relevant to the development and solution of the problems, rather than to introduce a problem that was of importance in some previous age and to trace the development of that problem down the years. Notwithstanding this approach, however, it remains true that many of the problems of space and time that are with us today were alive also to philosophers over two thousand years ago, so philosophers from Aristotle onwards will be entering the discussion as it proceeds.

The reader may find in places that he is being led on from point to point without quite knowing where the argument is taking him. This is in general a writing practice to be avoided whenever possible. It is preferable to be able to state the problem succinctly, to state the conclusions one is going to draw and how one is going to go about it, and then proceed to do just that. This is fine given that there is a common enough background between reader and author for the problem to be stated succinctly, but often this is not the case. It is the prime object of this book to provide that background. Besides, 90 per cent of the difficulty with many a problem, particularly the problems presented here, is to be able to state the problem succinctly and unambiguously. All this is not to say that no attempt is made in the book to solve the problems thus raised. Further, the attempt is made throughout to keep the reader as informed as possible as to what is going on, given the assumption that the reader is being introduced to the topics.

I wish to acknowledge my debt to J. J. C. Smart and C. B. Martin who first stimulated my interest in the philosophy of space and time, and to Michael Bradley and Angus Hurst for their encouragement at that time. I would also like to thank the members of the Department of Philosophy at the University of Michigan who invited me to lecture on this topic in the northern winter of 1967. This book is largely a development of those lectures. The debt to Adolf Grünbaum and his book *Philosophical Problems of Space and Time* will be evident from the many references made to Grünbaum and his book throughout the text. Thanks are due also to Malcolm Rennie, Gary Malinas, Roger Lamb, and John Briton of the University of Queensland, J. J. C. Smart of La Trobe University, Rom Harré, and Michael Hinton of Oxford University and John Bennet, Arthur Burks, Larry Sklars, and Jack Meiland of the University of Michigan, all of whom read some of the manuscript and offered very helpful advice and criticism and, above all, encouragement. Finally, I thank Professor D. J. O'Connor of the University of Exeter, who first suggested that I write this book.

Contents

5. EXISTENCE AND THE PRESENT

6. TEMPORAL ASYMMETRY

List of Figures

Introduction

As a rule, people differentiate between matter, space, and time. Matter is what exists in space and endures through time. But this does not tell us what space is, and it does not tell us in what way time differs from space. It is matter which we see, touch, and hear, which causes sensations to arise within us, and which is causally operative on other matter. But if space is not causally operative on anything, including ourselves, how can we know that it is there? If it is causally operative, how can we distinguish it from matter? Is space just a special kind of matter with physical properties of its own? But if it is, can there be such a thing as *empty* space? Similar questions arise with respect to time, but time presents difficulties of its own. Time passes; but what is this thing or stuff that is passing and what is it passing? If it is flowing past in some sense, does it make sense to ask how fast it is flowing? Perhaps none of these questions makes sense, or perhaps false presuppositions lie behind these questions. But if so what are these false presuppositions? Why are they false? Why, if it is so, do the questions not make sense?

These are the basic problems with which this book is concerned. Roughly speaking Chapters 1, 2, and 3 deal with space and Chapters 4, 5, and 6 deal with time. I say roughly speaking, for as we shall see the problems are intertwined.

Of course none of these problems would arise if we never felt the need to refer to space or time. Perhaps these problems arise merely by virtue of the way in which our language is constructed. Perhaps the problems are pseudo-problems to be resolved by using a reformed language more appropriate in describing this world. Or, at least, perhaps whatever truth lies in statements, which seem to refer to space and time, can be expressed in statements which make no such reference. The belief that this is so is referred to throughout the book as the relational theory of space and time. The belief that we cannot without loss drop reference to space and time is called the absolute theory of space and time. This terminology is explained in more detail in sections 2 and 3 of Chapter 1.

The business of providing means for the elimination of certain expressions from our descriptions of the world is often called reduction. Thus relationalism is a reductionist programme. But there are many different kinds of reduction and confusion often arises as the result of a failure to distinguish between the different kinds of reduction. Section 4 of Chapter 1 draws distinctions between four different kinds of reduction.

In Chapters 2 and 3, many different properties of space are investigated with a view to seeing whether the relationalist programme can be made to work with respect to these properties. In some cases, the reduction is quite easily managed; with others, difficulties are encountered.

In Chapter 4 the differences between space and time are investigated. Some seemingly obvious differences disappear on closer inspection. Chapters 5 and 6 are devoted to an investigation of one of the more important of these 'differences'—the belief that time flows and that space does not.

Finally, in the Conclusion it is argued that the relationalist programme can always be made to work, but in some cases only by the invention of a surrogate entity to bear the properties that prior to the reduction were alleged to be borne by space or time. In such cases, either acceptance of the relationalist reduction or the acceptance of an absolute space or time will as like as not be accompanied by the need to reject some of our fundamental presuppositions about the nature of the universe.

CHAPTER 1

Space—Relational or Absolute?

1.1 WHAT IS SPACE?

Why is it that the question 'What is space?' is so difficult to answer?
It is not that the word 'space' is unfamiliar to us. For example, the
difficulty is not the sort of difficulty that most of us would have in
answering the question, 'What is pyruvic acid?' The great majority
of people have not heard of pyruvic acid. But surely almost everyone
has heard of space and, furthermore, often makes reference to space
in such sentences as:
> 'There's not much space here.'
> 'This table takes up a lot of space.'
and 'The astronauts are floating around in space.'

However, specialized knowledge *is* sometimes required to answer
a 'What is . . .' question, even if the noun we insert into the blank
is in common usage. Consider for example:
> 'What is electricity?'
> 'What is soap?'
> 'What is D.D.T.?'
There is a level at which such questions are answerable without
much expertise, for example as follows:
> 'That's what makes the light glow.'
> 'That's what we wash with.'
> 'That's the stuff in the flyspray that kills the insects.'
It is when it is clear that such facts are known that expertise is
required. But, on the surface of it, it would seem rather absurd to be
rushing off to physicists and chemists to find out what they believe
space to be, and this for two reasons. Firstly, we might feel that in
spite of our difficulty in answering the question 'What is space?' it is
nevertheless the case that, if anyone knows what space is, surely we
do. Secondly, we might feel that space, unlike pyruvic acid, electricity, soap, and D.D.T., is not a substance, let alone a substance
whose composition is unknown to us.

Nevertheless, the way we so often describe space is just as if it is a
substance. Compare the examples above with the following:
> 'There's not much water here.'
> 'This sponge takes up a lot of water.'
> 'The dust particles are floating around in the water.'

Why, then, do we feel that space is *not* some sort of substance? There are such facts as that we can either absorb, breathe, drink, or eat substances, if the portion is small enough, depending on whether the substance is radiation, gas, liquid, or solid. But space does not seem to be the sort of thing we can absorb, breathe, drink, or eat. We may think it bizarre that there is a man in Western Australia who eats pieces of motor-car as a fair-ground entertainment, but it seems to be logically incorrect to talk of a man consuming 100 cubic centimetres of space for lunch.

The question arises as to whether or not we could generalize on this example. It makes sense to say that a piece of some substance reacts with or interacts with another piece of some substance. Does it make sense to say that some bit of space reacts with or interacts with a piece of some substance? Does it make sense to say that some bit of space reacts with or interacts with another bit of space? Many would feel that the answer to these questions is no: space is simply that in which reactions and interactions may take place, without itself taking part in the proceedings. It necessarily remains neutral and passive throughout.

But if space never interacts with any piece of matter, then it would never interact with any observer, even indirectly; that is, via a piece or some pieces of matter that themselves interact with an observer. This being the case, how is it that anybody has come to believe in the existence of space? Why do people continue to believe in the existence of space? Is it just a piece of mythology handed down to us from our ancestors which, unlike the myths about gods, ghosts, and fairies, we have never stopped to query? But this seems absurd. Surely the existence of space is immediately evident to anyone who uses his eyes—or for that matter his hands. Where then have we gone wrong?

1.2 RELATIONAL THEORIES OF SPACE

Some would say that where we have gone wrong is that we have allowed the grammar of the word 'space' to mislead us. Thus, because 'space' is a noun, we have been misled into asking such questions as 'What is space?' and 'Can space interact with other things?' in much the same way as we might ask 'What is bread?' or 'Can Phosphorus react with Nickel?' Such questions have led us on to the strange question 'Does space exist?'

'But', it may be added, 'it is not because the answer is obviously "Yes" that the question is strange. It is strange because the question seems to presuppose that space is some sort of thing or substance. What there are in the Universe', the claim continues, 'are pieces of matter composed of various substances and these pieces of matter exhibit spatial relationships between each other and between their own parts. Any statements purporting to assign any property or properties to space are to be reconstrued as assigning relationships to pairs or groups of pieces of matter.' This sort of view is called a *relational* theory of space. Relational theories of space and time have been under consideration for many hundreds of years, dating back at least as far as the ancient Greeks.

Some relational theories of space treat the logic of 'space' much as we treat the logic of 'friendship' at least in the following respect. To say that 'friendship' exists is just to say that someone is a friend of someone else. To say that space exists, so goes the story, is just to say that something is apart or at a distance from, something else. We don't have to say that it is always the case that something, namely space, lies between two objects placed at a distance from one another. Sometimes, of course, we shall want to say that some substance, say water or air, lies between the two objects, but sometimes we shall want to say that nothing lies between the two objects.

1.3 ABSOLUTE THEORIES OF SPACE

As opposed to the view mentioned in the previous section, there is the view that descriptions of the space between objects cannot, without loss, be equated to descriptions of relations obtaining between those objects. To describe space as a substance may or may not be a bit too much, but certainly, so goes this view, space is an entity in its own right with properties of its own. Empty space may be space which contains no substance, but it is not nothing.

What sort of properties does the absolutist allow space to have? Of course the answer to this question could vary from one absolutist to another. So long as one allows that there is at least one property that space has which cannot be translated away or otherwise 'reduced' to a relation between objects, one is an absolutist. Thus, whenever an absolutist mentions a property which space has, there is a challenge to the relationalist to reduce the theory that space has this property to an equivalent theory in which space is not mentioned.

For example, consider the statement that empty space is transparent (just as water and glass are transparent). What can the relationalist make of this? He might say that to state that space is transparent is just to state that if there is nothing in the path of a light ray the light proceeds unimpeded. This reduction does seem to be equivalent to the original statement that empty space is transparent, but it no longer seems to refer to some entity called 'space'.

1.4 WHAT IS A REDUCTION?

I have just described the above example as a reduction and in that context it was probably clear enough, for the purposes of the example, what was meant. However, since the aim of the relationalist is always to reduce the absolutist's descriptions of space to descriptions of relationships between other entities, it might be an idea to look a little more closely at what reduction can amount to.

'Reduction', in the sense of the word as used by philosophers, is a technical word, yet one whose usage is not cut and dried. One thing that can be said fairly definitely, however, is that in philosophic discourse 'reduction' is used to describe the replacement of one manner of speech by another, such that the new manner of speech does not use some word, term, or expression that the original manner of speech used. Thus in the example above, a statement in which space was described as transparent was replaced by a statement which made mention only of unimpeded light rays, and in which the word 'space' did not feature.

However, this criterion is not sufficient to establish that a reduction in this sense has taken place. No philosopher would believe that 'The cat is on the mat' could be reduced to 'The dog is in his kennel' despite the fact that 'cat' and 'mat' do not feature in statements concerning only dogs and kennels. What else, then, should be added?

Some might argue that the difference in the two examples mentioned is that the first case (the reduction of the description of space as transparent to a tautology concerning unimpeded light) was a case of synonymy, whereas the second case about the cat and the mat and the dog and his kennel was not. Certainly what many philosophers seem to have in mind when they describe something as a reduction is often a case of synonymy. For example, by virtue of the fact that 'Neither p is the case nor q is the case' (where some sentential expressions that are neither questions nor commands may be substituted for p or for q) is synonymous with 'It's not the case that p and it's not the case that q', many philosophers would say that expressions involving 'neither . . . nor . . . can be reduced to expressions containing 'it is not the case that . . .' and 'and', and not containing 'neither . . . nor . . .'

Similarly, one might say that 'bachelors' could be reduced to 'unmarried marriageable males'. However, 'is a reduction of' in the sense of 'has the same meaning as' is not often used in cases where it is perfectly well known that the two expressions are synonymous. The word 'reduction' is usually reserved in such contexts to those cases where knowledge of the synonymy relation comes somewhat as a surprise—where, perhaps, argument has to be used in order to show that the two expressions are synonymous.

Often there is a prescriptive force in the use of the word 'reduction' by philosophers—even where it is meant that the statement to be reduced is synonymous with the reducing statement. How can this be so? If the two statements do in fact have the same meaning, why should anyone bother whether one is used rather than the other? One reason is that some expressions can be (as Gilbert Ryle has described them) systematically misleading. That is, because of the grammar of the statement in which the expression is couched, one may be misled into making unwarranted inferences from the statement. Thus on page 4 we had an imaginary philosopher saying that we had been misled (by the fact that the word 'space' was a noun) into thinking that 'space' referred to some thing or substance. To overcome such mistakes (if they are mistakes) someone might feel like encouraging people to use the synonymous expressions or statements which do not exhibit the misleading grammar. So the *first* sense of the word 'reduction' involves ridding a statement or class of statements of some term or expression and is such that the reducing statements are alleged to be synonymous with the statements to be reduced.

I have identified two areas where reduction by synonyms is often of importance. In the first area the fact that the statment to be reduced is synonymous with the reducing statement is noteworthy in itself. In the second area, the synonymy between the two statements may or may not come as a surprise, but the grammar of the reducing statement is less logically misleading than the grammar of the statement to be reduced, and it is for this reason that the reduction is pointed out. For example, Gilbert Ryle has claimed that 'Colour involves extension' means what is meant by 'Whatever is coloured is extended', but that whereas the second expression does not seem to refer to something named 'Colour', the first does seem to do so, and Ryle claims that there is no object named 'Colour'. I shall henceforth call a reduction which is based upon synonymity, and which is carried out for either of the above reasons, an *analytic* reduction.

The second type of reduction is one in which—like the third and fourth types, yet to be explained—no claim for synonymy is made for the two statements concerned. In the second case, what is claimed is that the statement to be reduced is inexplicit—it is difficult if not impossible to know what implications one may draw from the statement. The reducing statement does some, or most, of the job that the reduced statement does, but is more explicit. Often such reductions are called 'explications' or 'rational reconstructions'. I shall henceforth call them *'explicatory reductions'* or 'explications' for short. As an example of this philosophers and logicians who have despaired of finding a logic for the words 'if . . . then', which seems to suit all contexts, often resort to an explication of 'If p then q' in terms of 'Either it's not the case that p, or q is the case', a statement form whose logic is well understood, and which generates a large proportion of the implications which are generated by 'If . . . then'. The following would exemplify a typical attitude of a reductionist of the second kind:

'Goodness knows what people mean to imply when they seem to be describing something they call space. Some seem to want to imply the existence of a some absolutist thing or entity—others do not. All, however, seem to wish to refer to spatial relationships between bodies or parts of bodies, so let us see if we can manage just with references to spatial relationships alone. This would at least be more explicit.' Thus with the second category of reductionists also there is a prescriptive aspect.

The third type of reductionist is one who believes that the statement to be reduced *does* have implications which the reducing statement does not have. The expression which is present in the statement purports to refer to an entity which, so he claims, does not in fact exist. Since explanations which presuppose falsehoods are liable to be misleading, the reducing statement, which has all the implications of the original statement with the exception of those asserting the existence of the entity in question, has all the useful explanatory power of the original statement. Thus, to return to our example of the transparency of empty space, such a reductionist might claim that the statement that empty space is transparent does indeed imply the existence of some thing or substance called 'space' as the absolutist would claim, but where the absolutist is wrong, so the claim would continue, is that he believes that this thing or substance exists. A reduction of this kind will henceforth be called an *ontological* reduction.

Closely allied with this position is one which, beginning with the assumptions that both reducing statement and the statement to be reduced have the same explanatory power, and that the statement to be reduced implies the existence of a certain entity or entities whose existence is not implied by the reducing statement, proceeds to draw the conclusion that it would be irrational to believe in the existence of the entity or entities in question. Such an attitude derives from a dictum which is widely used (and also criticized) in many areas of philosophy, and which is called Occam's Razor. The dictum is: 'Do not multiply entities beyond necessity.'

The fourth type of reductionist is best illustrated by an example.

We are told by physicists that all light is electromagnetic radiation with a wavelength lying between 4000 and 7000 ångströms (Å). It is not that 'light' means 'electromagnetic radiation with a wavelength lying between 4000 and 7000 Å', it is just that light *happens to be* electromagnetic radiation of that wavelength and that this is something that our scientists have discovered about the world.

To introduce some terminology: when property *A happens to be* the same property as property *B*, we say that property *A* and property *B* are *contingently identical*. But when property *A* and property *B* are the same property by virtue only of the meanings of the expressions used to designate property *A* and property *B* we say that property *A* and property *B* are *necessarily identical*. Thus if 'bachelor' is synonymous with 'unmarried marriageable male' then the property of being a bachelor is necessarily identical with the property of being an unmarried marriageable male.

On the other hand, 'light' is not synonymous with 'electromagnetic radiation with a wavelength lying between 4000 and 7000 Å' so, in so far as the property of being light *is* the property of being electromagnetic radiation with a wavelength lying between 4000 and 7000 Å, the property of being light is contingently identical with the property of being electromagnetic radiation with a wavelength lying between 4000 and 7000 Å. Or, to put it in another way (a very ambiguous way as we have seen), 'light' can be reduced to 'electromagnetic radiation with a wavelength lying between 4000 and 7000 Å'.

Could such a reduction apply to space? There has been at least one physicist, E. J. Zimmerman, who has made such a suggestion. In order to describe what we normally think of as sub-microscopic happenings, we have increasingly to rely on that branch of physics called quantum mechanics. But just as the idea of a temperature is inapplicable to the items of the molecular theory of heat, namely the molecules, so it is, Zimmerman claims, that the ideas of spatial and temporal position are inapplicable to the items of quantum mechanics. Nevertheless it may be the case that spatial and temporal properties, needed in our descriptions of macroscopic effects, are reducible to statistical effects of the properties of these quantum mechanical items—properties such as charge, mass, strangeness, and spin. If this doesn't sound very commonsensical then, as J. J. C. Smart has pointed out, such an outcome could hardly be surprising; for commonsense notions are the notions of macroscopic phenomena —not of sub-microscopic events.

But the point to be made here does not concern the question of whether or not Zimmerman's conjectures make sense, let alone whether or not they are correct. The point is that they are attempts at a theoretical reduction of space and spatial properties. Let us recapitulate. Four different types of reduction have been mentioned.

Reductions of the first type were called *analytic* reductions— 'analytic' because the reduction purports to provide analyses of the meanings of the corresponding descriptions of space.

Reductions of the second type were called *explicatory* reductions— 'explicatory' because the reductions purport to provide an explication of the corresponding descriptions of space or, to put it another way, the reductions purport to give us a way of describing the world just as richly as those ways involving descriptions of space allow, but in a more explicit manner.

Reductions of the third type were called '*ontological*' reductions. 'Ontological' means 'pertaining to existential beliefs'. The point of ontological reductions is to ensure that we do not go about saying things which have false implications concerning the existence of some entity, just for want of some other way of expressing ourselves.

Finally, reductions of the fourth type (where, for example, someone says that space is contingently identical to something or other) were called theoretical reductions.

Relational theories of space are not so much theories of space *per se*, but rather affirmations of a belief that a programme which has as its aim the reduction of statements describing space could be successful. Likewise we could regard an absolutist theory of space as one which entails the belief that no such reduction would be successful. In the Conclusion it is argued that the relationalist programme can always be made to work in any particular case, but some cases could require drastic changes in our ideas of matter.

A final note of warning before leaving this section on reduction. The reader may have been misled into thinking that any one philosopher's attempt at bringing about a reduction could be categorized into one of the four categories listed above. Often, however, it is very difficult to see what sort of reduction such a philosopher, be he Theophrastus, Berkeley, or Mach, is trying to bring about. Conceivably, also, it would be possible for a theory which a physicist or philosopher is propounding to be an exemplification of two or more of these types of reduction at the same time. Take for example Maxwell's molecular theory of heat. In the eighteenth-century sense of the word 'heat', namely the sense in which heat was regarded as a sort of fluid, Maxwell's theory of heat is not a theory of heat at all. Part of the point of Maxwell's theory of heat was to spell out that there is no such fluid. Thus Maxwell's theory of heat changed the meaning of the word 'heat' to something of greater explanatory use, and in this sense provided an explication of the word. At the same time the reduction could be called 'ontological' because the old notion of heat as a fluid was no longer required— there being no such stuff according to Maxwell. Lastly, but not least, the new theory gave us a theoretical reduction, in so far as it yielded a contingent identity between the property of being hot and the property of having a high average molecular kinetic energy.

1.5 SOURCES AND HISTORICAL NOTES FOR CHAPTER 1

Problems of the sort introduced in section 1 were being considered by philosophers in Greece more than two thousand years ago. Archytas distinguished space from matter, yet nevertheless endowed space with a considerable number of substantial properties such as pressure and tension. Atomists such as Democritus opposed this view, requiring the space between their atoms to be empty extension without any causal properties at all. Melissus argued that a vacuum was impossible, using an argument that was repeated by Descartes thousands of years later and which will be discussed in some detail in Chapter 3, sections 3.5 and 3.6. Relational theories of space, presumably designed in order to avoid such conclusions, were enunciated in those times, for example by Theophrastus, a pupil of Aristotle.

These origins are outlined in Chapter 1 of Max Jammer's *Concepts of Space* (Harvard University Press, Cambridge, Mass., 1954). The chapter entitled 'The Concept of Space in Antiquity' is reprinted in the reader *Problems of Space and Time*, edited by J. J. C. Smart (Macmillan, New York, 1964). Jammer's article includes an extensive bibliography.

Among the better-known philosophers of more recent times, René Descartes and Isaac Newton are perhaps the most famous absolutists. Isaac Newton's theories of space and time have been the basis for the study of dynamics from the seventeenth until the present century. Gottfried Leibniz, Bishop Berkeley, Ernst Mach, Albert Einstein, Henri Poincaré, and Adolf Grünbaum are among those who have argued against the Newtonian concept of space and the need to set physics on a relationalist footing. It is sometimes misleading to classify philosophers in this way, however, for the views of any two relationalists on the subject of space can be very different—while on some aspects of the subject a relationalist and an absolutist can have similar if not identical views. Thus the absolutist Descartes and the relationalist Leibniz had in common the view just attributed to Melissus—that there is no such thing as a vacuum.

Descartes's views on space occur in his *Principles of Philosophy*, Part II, and English translation of which is to be found in *Descartes:*

Philosophical Writings, edited and translated by Elizabeth Anscombe and Peter Geach (Thomas Nelson and Sons, Edinburgh, 1954). Excerpts from this are reprinted in J. J. C. Smart's reader mentioned earlier. This reader also contains appropriate selections on the subject from the works of Isaac Newton, Gottfried Leibniz, Ernst Mach, Albert Einstein, Adolf Grünbaum, and other notable authors in the field.

Bishop Berkeley's criticisms of Newton's views on space are to be found in *The Principles of Human Knowledge*, paragraphs 111–17, available in *Berkeley. A new theory of Vision and other writings* (Everyman's Library, J. M. Dent and Sons, Ltd., London, 1910). Henri Poincaré's philosophically exciting defence of relationalism is to be found in his *The Foundations of Science*, translated by G. B. Halstead (Random House, New York, 1970). A very clear account of the absolutism versus relationalism debate which takes its departure from Newton's views and Leibniz's criticisms thereof is to be found in Chapter 4, section 1 of Bas van Fraassen's excellent text *An Introduction to the Philosophy of Time and Space* (Random House, New York, 1970). The discussion of absolutism and relationalism in van Fraassen's book and also in Adolf Grünbaum's book *Philosophical Problems of Space and Time* (Routledge & Kegan Paul, London, 1964), is primarily a discussion of and a development of the debate between the Newtonians and the anti-Newtonians. In this book I try to put the problems in as broad a philosophical setting as I can with the result that my usage of the terms 'absolute' and 'relational' will probably give these terms a wider application than is common. If I have thus stretched philosophic usage a little, my excuse is that I hope that the philosophical boundaries thus delineated will be simpler, generally more useful, and less dependant on the idiosyncrasies of a particular period in the history of philosophy.

With respect to reduction, philosophers have been reducing statements for thousands of years, but as far as I know, they have been describing themselves as doing this for only several decades. It may be that other attempts to classify and compare the different kinds of reductions are about, but I do not know of any. Gilbert Ryle's example involving the reduction of 'Colour involves extension' comes from his article 'Systematically Misleading Expressions' in *Logic and Language* (First Series) edited by A. G. N. Flew (Basil Blackwell, Oxford, 1953). It was Max Deutscher, in a paper here read to the Annual Conference of the Australasian Association of Philosophy in 1964, who first convinced me that properties could be contingently identical. The paper is published in a book entitled *The Identity Theory of Mind*, edited by C. F. Presley (University of Queensland Press, Brisbane, 1967).

CHAPTER 2

The Properties of Space

2.1 INTRODUCTION

In section 1.3 it was mentioned that the absolutist regards space as an entity in its own right and with properties of its own. Furthermore, whether the relationalist likes it or not, scientists and laymen alike attribute many different sorts of properties to space. The list that follows is a fairly representative sample. The reader is not to be alarmed if he does not at this stage understand all the terminology which appears in the examples. The lists under the various headings are by no means meant to be complete. It is not being alleged, of course, that each item in the lists is a true statement, but only that it has been believed to be true by someone at some time.

(a) *Electrical, optical, and electromagnetic properties of space*
 (i) Empty space is a poor conductor.
 (ii) The magnetic permeability of empty space is $4\pi \times 10^{-7}$ henrys per metre.
 (iii) The permittivity of empty space is $8\cdot55 \times 10^{-12}$ farads per metre.
 (iv) The speed of light in empty space is $2\cdot9978 \times 10^8$ metres per second.
 (v) Empty space is transparent.

(b) *Kinematic and dynamic properties of space*
 (vi) When any body moves with respect to another, at least one of the bodies is moving with respect to absolute space.
 (vii) The sum of the external forces upon a body is other than zero if and only if the body is accelerating with respect to absolute space.
 (viii) Space is penetrable.
 (ix) Space is incapable of action.
 (x) The parts of space cannot be separated from one another by any force, however great.
 (xi) Space is immovably fixed.

(c) *Metrical, topological, and geometrical properties of space*
 (xii) Space is a continuum of infinitesimal points.
 (xiii) Space is infinite.
 (xiv) Space is finite but unbounded.

(xv) Given any three points in space, A, B, and C, such that the angle ABC is a right angle, then the square of the distance AC is equal to the sum of the squares of the distances AB and AC.

(xvi) Space has a non-Euclidean geometry.

(xvii) Space is curved.

(xviii) There is a finite distance between any two points in space whose magnitude depends only on the two points selected.

(xix) Space is three-dimensional.

(d) *Miscellaneous*

(xx) All bodies are placed in space.

(xxi) Each point in space has associated with it at any one time a gravitational field strength, an electric field strength, and a magnetic field strength.

(xxii) Space is isotropic.

In so far as he believes any one of these statements to be true, the relationalist must attempt to reduce it in one of the senses explained in section 1.4, to a statement which does not use the term 'space'. The sections which follow in this chapter and the next, attempt to demonstrate the kind of difficulties to be met in such an attempt.

2.2 THE CONDUCTIVITY OF SPACE

Consider the following experiment. The apparatus consists of a fully charged 12-volt car battery, some copper wire, some glass, an ammeter, with a range of one amp, and a 12-volt, 12-watt filament lamp. The copper wire and the ammeter are connected with the battery and the lamp as shown in Fig. 1.

We note that the lamp shines brightly and that the ammeter shows that about one amp of current is flowing.

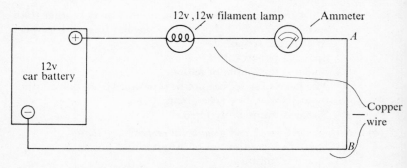

FIG. 1. *Experiment indicating high conductivity of copper wire*

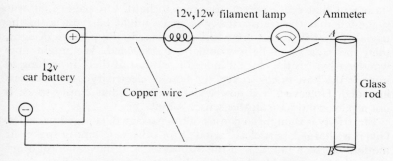

Fig. 2. *Experiment indicating low conductivity of glass rod*

We now remove the copper wire between points A and B and replace it with our glass rod as shown in Fig. 2.

We note now that the lamp no longer shines and that the ammeter indicates no current flowing. Now we remove the glass rod so that points A and B are joined only by air. Once again no current flow is indicated. It is by virtue of the results of such experiments that copper is said to be a good conductor of electricity and that glass and air are said to be poor conductors of electricity.

What would happen if we had literally nothing between A and B; that is, if the experiment was conducted in a vacuum as in Fig. 3?

Once again the ammeter would indicate no current flow. Are we then entitled to say that empty space is a poor conductor of electricity? A relationalist might say 'O.K., so long as you do not mind my reducing this statement to some such statement as the following:

"If there is nothing to conduct electricity, then no electricity is conducted."'

But one may be forgiven for regarding this reduction with some suspicion, especially in view of the method of reduction used for the

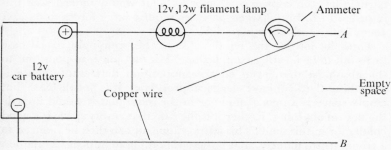

Fig. 3. *Experiment indicating low conductivity of empty space*

statement that space was transparent to light. On page 5, it was suggested that the transparency of space might be reduced to the triviality that if there were nothing to impede the passage of light, then the passage of light would be unimpeded. We might now wonder why it is the case (if indeed it is) that, if there is nothing to impede the flow of electricity, the flow of electricity is nevertheless impeded. However, it is now widely believed that empty space is not a poor conductor in this sense. Why?

Electricity is thought to consist of the passage of charged particles. Given a charged particle in what is an otherwise empty space, its motion will be a function of the disposition and movement of other charged bodies outside that space, but that is all.

Within these constraints, charges in empty space will move quite unimpeded. If they are negatively charged they will be repelled from other negatively charged bodies and be attracted towards positively charged bodies and *mutatis mutandis* for positively charged bodies.

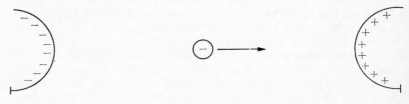

FIG. 4. *Charged bodies move quite freely in empty space*

How can these beliefs be reconciled with the results of the sort of experiments described earlier in this section? It is thought that the impedance to the current flow in the experiment shown in Fig. 3 is not in the empty space, but rather in the fact that charged particles do not leave a conducting material readily. If a negatively charged particle left the copper conductor at A, it would indeed pass unimpeded to B, but charged particles require large forces to extract them from conducting materials.

Thus the job for the relationalist in this seemingly trivial case turns out to be a rather complex one. His first job is to point out to the absolutist that, even on the supposition that empty space is something that can have electrical properties, it is not the case that empty space is a poor conductor in the sense that it would impede the flow of anything. Rather it is the case that empty space is a good conductor in this sense. This latter statement can then be reduced to a trivial truth, namely, that if there is nothing to impede the motion of a charged particle then the motion of the charged particle is unimpeded. Empty space is a poor conductor only in the sense that a dry river bed is a bad place to look for a current of water. Empty space has within it no charges whose motion would be a current.

The results of the experiment of Figs. 1, 2, and 3 then remain to be explained, and this is done by virtue of theories concerning certain relationships that charged bodies bear to conductors and which do not make mention of empty space.

So far, so good for the relationalist. But what we have seen in this example, is that the relationalist cannot always opt for similar solutions given different properties of space. Also the question 'Why?' readily arises when the relationalist opts for different types of solution. It is in the furnishing of this explanatory detail in particular cases that the difficulty for the relationalist lies. But the search for such explanatory detail can be a stimulant for research in physics.

2.3 THE PERMITTIVITY OF EMPTY SPACE

In this section, a relationalist reduction of the statement that space has a permittivity of 8.55×10^{-12} farads per metre is developed. But in section 2.4 it is shown that this reduction is unsatisfactory when paired with a similar reduction for the statement that space has a magnetic permeability of $4\pi \times 10^{-7}$ henrys per metre. First of all, however, the notion of 'permittivity' is explained. The notion of 'magnetic permeability' is explained in section 2.4. The reader is warned that the experiments described in these sections are not recommended as the best way to make the measurements involved. The descriptions of these experiments are primarily intended to introduce the concepts of 'permittivity' and 'permeability'.

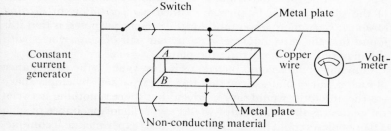

FIG. 5. *Measurement of capacitance of parallel metal plates separated by non-conducting medium*

Consider the apparatus depicted in Fig. 5. The apparatus consists of a source of electrical current which, as long as the switch is closed, emits a constant electrical current, of a certain number of amps, say I amps. A and B are two parallel plates made of some conducting material such as copper. The plates are separated by some non-conducting material such as glass. A voltmeter measures the voltage across the plates. For simplicity, let us imagine the voltmeter to be so constructed that only a negligible current passes through the voltmeter. Thus the current from the current source can pass only on to the plates. Some sort of timing device, connected to the switch, is used to measure the time (call it t) during which the switch is thrown.

Now the magnitude of an electrical current is, by definition, the amount of electrical charge flowing per second. Let us use Q to indicate the amount of charge which flows on to the plates during the time the switch is closed. Then since the current, I, is constant during the time the switch is closed, the amount of charge which flows on to the plates during this time is given by

$$Q = I\,t$$

Thus one can calculate the amount of charge on the plates and compare this with the voltage across the plates. One finds that the ratio of charge to voltage remains a constant.

That is, $\dfrac{Q}{V} = C$, where C is a constant.

If the experiment is repeated, varying the distance between the plates, the area of the plates, and the insulating material between the plates, it is found that the value of C varies. For example, if the dimensions of the plates are large compared with the distance between them, then the value of C is proportional to the area of the plates and inversely proportional to the distance between the plates. The value of C for a particular configuration is called the *capacitance* of the plates, which in turn are called a *capacitor* or an *electrical condenser*. Even more important, for the present purposes, is that it is found that the capacitance of a particular pair of plates varies with the material used to separate the plates.

One thing which could interest us is this. What is the capacitance of such a pair of plates given that they are separated by empty space (as the absolutist would say) or given that there is nothing between the plates (as the relationalist would say)? The capacitance of such a pair of plates could be found using the same experimental technique.

We have said that the capacitance of a parallel plate condenser is proportional to the area of the plates and inversely proportional to the distance between the plates (so long as the distance between the plates is small compared with the other dimensions). Let A be the area of the plates of such a condenser and d the distance between them. Let ϵ be the constant of proportionality for the particular substance separating the plates. Then we may write:

$$C = \frac{\epsilon A}{d}$$

Given, then, that we have measured C, A, and d, we may calculate the constant ϵ for any given material, and the value of ϵ for any given material will be an important property of that material. It is called the *permittivity* of the material. Since capacitance is measured in farads, then if A and d are measured in square metres and metres respectively, permittivity will be measured in farads per metre.

It is found that the permittivity for empty space (usually called ϵ_0) is $8\cdot854 \times 10^{-12}$ farads per metre. This does not seem to be the sort of result that can be translated away into something trivial. What can the relationalist do with this?

'Well,' he might say, 'all that has been shown is that the capacitance of a parallel plate condenser is modified by the presence of some insulating material lying between the plates. Were we to choose our unit of charge differently, we could make the value of ϵ_0 equal to unity. The value of ϵ for any particular material would then be merely the extent to which that material affected the capacitance of any capacitor. The choice of unity for the value of ϵ for so-called "empty space" would be appropriate, for then the statement that the permittivity of empty space is unity could be reduced to the triviality that if there is nothing between the plates of a capacitor, then the capacitance of that capacitor is unaffected.'

But this reply is not wholly satisfactory as we shall see in the next section.

2.4 THE MAGNETIC PERMEABILITY OF EMPTY SPACE

In most home television-receivers is a device which generates what is known as a saw-tooth current waveform. That is to say, the device puts out a current whose magnitude, when plotted against time, looks like the teeth of a wood saw (see Fig. 6(a)). The output current of the saw-tooth current generator in the television-receiver flows into a coil of copper wire, which, as a result, produces a varying magnetic field which deflects the electron beam in the picture tube from one side of the tube to the other.

Let us imagine an experiment set up with such a saw-tooth current generator and such a coil. The two are connected as shown in Fig. 7. We can imagine for the purposes of the experiment that the change in current from the saw-tooth current generator is slow enough for the voltmeter to respond to any changes in voltage across the coil. Alternatively, those readers familiar with the use of cathode-ray

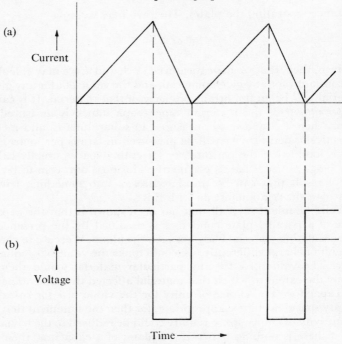

FIG. 6. (a) *A graph of the magnitude of the current emitted by a saw-tooth current generator, plotted against time*

(b) *A graph of the voltage across a coil through which the current in graph* (a) *is made to pass, plotted against the same time-scale*

oscilloscopes will know how such devices may be used instead of voltmeters for voltage measurements on rapidly changing waveforms.

In any case, given that the current flowing through the coil from the generator varies with time as shown in Fig. 6(*a*), then the corresponding changes in voltage which will be observed across the coil will turn out to be as shown in Fig. 6(*b*). A small positive voltage

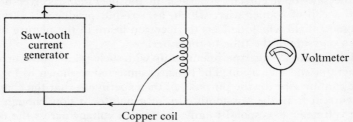

FIG. 7. *Measurement of inductance of a copper coil*

appears during those intervals in which the current is increasing slowly, and a large negative voltage appears during those intervals in which the current is decreasing swiftly. If we vary the rate at which the current increases, we find that the voltage across the coil varies proportionately.

Let the constant of proportionality be L. Let V be the voltage across the coil and let us express the rate of increase of the current I with respect to time with the symbol $\frac{dI}{dt}$. Then we have $V = L\frac{dI}{dt}$. L is called the *self-inductance of the coil*. Now the value of L varies depending on what the substance is, in which the coil is immersed. It is found also that, given that the length of the coil is great compared with its diameter, the self-inductance of the coil is proportional to the square of the number of loops per unit length, n, the area of the cross-section of the coil $(=\pi r^2$ where r is the radius) and is directly proportional to the length of the coil, l.

If we let the constant of proportionality be μ, then

$$L = \mu n^2 \pi r^2 l$$

μ is called the magnetic permeability of the material in which the coil is immersed. The value of μ for empty space (usually called μ_0) is exactly $4\pi \times 10^{-7}$ henrys per metre.

'What an extraordinary coincidence that it should turn out to be such a round fraction of 4π', someone might say. But, of course, it is no coincidence at all, for what physicists have done is to choose the value of μ_0 for empty space in order to *define* their unit of current. That is, with the magnetic permeability of empty space, they have done what the relationalist in the previous paragraph was asking them to do with the permittivity of empty space. But now having defined our unit of current, the permittivity of empty space becomes a matter which is to be found out by experiment—not by arbitrary assignation of values.

We could, of course, have defined the permittivity of space as unity or some other convenient value and thereby defined our unit of current, but then the value of the magnetic permeability of empty space would be something that could be determined experimentally.

To summarize, the point of the last two paragraphs is this. Any substance has associated with it a quantity called its permittivity and another quantity called its magnetic permeability. Both these quantities may be determined experimentally for any given substance. Space likewise has a permittivity and a magnetic permeability. The magnetic permeability of empty space is usually *defined* to be of a certain magnitude, the magnitude being chosen to give us a convenient system of electrical units based on the amp, the unit of electrical current. However, once this has been done, the permittivity of empty space becomes a matter for experiment. Here, then, seems to be one property of space that cannot be reduced away via some tautological triviality. Nor as the next section shows, can the statement that empty space has a permittivity of $8 \cdot 55 \times 10^{-12}$ farads per metre be reduced to a statement simply about the behaviour of charged particles on the plates of empty capacitors.

2.5 THE SPEED OF ELECTROMAGNETIC RADIATION IN EMPTY SPACE

The reason why statements about the permittivity of empty space cannot be reduced to statements about empty capacitors is that, even if there were no capacitors in the universe, the notion of the permittivity of a medium (and, by the way, the notion of the magnetic permeability of a medium) and the notion of the permittivity of empty space would still be needed to describe the behaviour of electromagnetic radiation.

In Maxwell's theory of electromagnetism, for example, the reciprocal of the square root of the product of the permittivity and the magnetic permeability of a medium turns out to be the speed of propagation of electromagnetic radiation through the medium. That is, if c is the speed of transmission of electromagnetic radiation in a medium, then the value of c is given by

$$c = 1/\sqrt{(\epsilon\mu)}$$

The value of c for empty space (call it c_0) is therefore

$$c_0 = 1/\sqrt{(\epsilon_0\mu_0)} = 2 \cdot 998 \times 10^8 \text{ metres per second.}$$

If Maxwell's theory is correct then electromagnetic radiation (which term, of course, encompasses radio waves, infra-red radiation, light, ultra-violet radiation, X-rays, and γ-rays) passes through space at some *particular* speed with respect to space.

'Well', someone might say, 'the speed of electromagnetic radiation in empty space would have to turn out to be some particular speed or other. For that matter, with respect to the last section, it is not surprising that the behaviour of electric charges on the plates of empty capacitors can be described with an equation using the universal constant ϵ_0. This constant would have to have some value or other. Perhaps we have been misled by the way we have extrapolated our use of the expression "permittivity of the medium" to the usage "permittivity of empty space". Why not restrict our use of the expression "permittivity of the medium" to those cases when there *is* a medium present and replace "permittivity of empty space" by "the permittivity when there is no medium present"? The fact that the capacitance of a capacitor is proportional to a permittivity would not then entail that that permittivity had to be a property of something or other. Likewise the fact that the speed of electromagnetic radiation bears some mathematical relationship to a permittivity does not entail that the permittivity has to be a property of a medium. The permittivity varies according to whether or not a medium is present, and *if* a medium is present it varies according to what sort of a medium it is. To allay confusion on this matter why not choose our unit of distance as well as our unit of electric current to make ϵ_0, μ_0, and hence c_0 all equal to unity? Any variations of permittivity, magnetic permeability, and the speed of light away from unity could then be said to be caused by the presence of some substantial medium.'

But this approach also has its difficulties if we accept Maxwell's laws for electromagnetic radiation. For what Maxwell's equations tell us is that $1/\sqrt{(\epsilon\mu)}$ is the speed of electromagnetic radiation *with respect to a medium*. But on a relationalist account, in the case of empty space, there is nothing there with respect to which something could have a speed! True, on a relationalist account, things could move with respect to one another in empty space, but the values of the speeds of something with respect to other things would surely *vary* if there was relative motion between those other things.

Thus, on a relationalist account, there does not seem room for a *particular* speed of light with respect to space. Maxwell's theory of electromagnetic radiation then seems, on the surface of it, to be inconsistent with a relationalist account of space. Is it? Before looking at this question, first let us look a little closer at a proposition that is more general than the proposition that space is a medium with respect to which electromagnetic radiation moves, namely the proposition that space is a medium with respect to which any object in motion moves.

2.6 KINEMATIC PROPERTIES OF SPACE

Does the sun move round the earth every day or does the sun remain at rest with the earth rotating on its axis once per day? As any twentieth-century schoolboy 'knows', it is the earth which rotates and the sun which remains at rest. But is there a difference between the two propositions?

Given two objects A and B and given the fact that B is moving with respect to A, it follows that A is moving with respect to B. Also if A is moving with respect to B, it follows that B is moving with respect to A. That is, 'A is moving with respect to B' and 'B is moving with respect to A' express the same proposition. It is not that if one is true then the other is false. Either they are *both* true or they are *both* false. So what is the point in arguing whether the sun is moving around the earth or whether the earth is rotating on its axis?

'But', someone may object, 'to know that A is moving with respect to B is only part of the story. From this we know that either A or B or both are moving, but we do not know that A is moving and we do not know that B is moving. All we know is that it is not the case that both are at rest.'

But what is it to say that something is at rest when we do not mean 'is at rest' to be taken as an ellipsis for 'is at rest with respect to the surface of the earth' or 'is at rest with respect to the sun' or 'is at rest with respect to some larger system'?

The ancients must have had some such notions of absolute rest and absolute motion, for various men from Anaxagoras to Bruno are alleged to have been at odds with their contemporaries in believing that the earth moved. For the ancient Greeks and medieval Christians alike, it was important to believe that the earth was immobile and that it was the sun, moon, planets, and stars which moved.

Can this attitude of the ancients be reconciled with the belief that if anything moves, it moves with respect to something else? If part of what we meant by absolute motion was that something could move absolutely, without moving with respect to anything else, the answer would of course, be 'no'. But what if we meant by the absolute motion of an object, motion that was independent of the motion of other objects? Is there a difference? That would depend on what we were willing to countenance as an object. If we were willing to countenance the existence of space and yet deny that space was an object, we could allow that absolute motion was motion with respect to space. Since nothing can move with respect to itself, space itself would be absolutely at rest. This was Newton's position with respect to space. For Newton, space was penetrable yet immovably fixed, and hence the parts of space could not be separated from one another 'by any force, however great', and in these respects, of course, it was most unlike anything we would call an object. Also space was that with respect to which objects could be in absolute rest or in absolute motion. Newton did not, of course, share the belief of the ancients that the earth was absolutely at rest.

Does a belief in absolute motion commit one to a belief in motion with respect to absolute space? On the surface of it, this would not seem to be so. Newton's idea of absolute motion as being motion with respect to space, was introduced as one way of reconciling the proposition that there can be motion independent of motion with respect to other objects, with the proposition that if anything moves it moves with respect to something. However it is conceivable that someone's notion of absolute motion was such that there could be something in absolute motion even though it was false that it was in motion relative to any other thing whatsoever, including space or any other thing which we may not care to call an object. Such a person would be one who would believe himself capable of conceiving of a universe in which everything moved with the same velocity in the same direction and in which, therefore, all relative motion was zero.

Alternatively, if one believed in the existence of objects that were at absolute rest, then if anything moved it would be moving with respect to such objects. For Aristotle, for example, something was at rest if it lacked an 'efficient cause' to keep it in motion. The notions of what we have called absolute rest and absolute motion had substance for Aristotle in so far as he believed that, if an object were causally isolated from other objects, it was at absolute rest. No two bodies in absolute rest, of course, were supposed to move with respect to each other. The important thing for us to note here is this. Assuming that some sense can be made of these positions, they are different positions from that of Newton. It was said that Newton's conception of space was one way of reconciling a belief in some sort of absolute motion with the proposition that if anything moves it moves with respect to something. Are there other ways of bringing about the reconciliation?

What if one said the following? Something is in absolute motion with a speed v if and only if it has a relative speed v with respect to anything whatsoever, regardless of that thing's motion relative to any third thing. In considering such a proposition, we are now, doubtless, far removed from anything the ancients said or thought; but let us not restrict ourselves to the cogitations of the ancients.

'But please', the exasperated reader may cry, 'let us at least restrict ourselves to that which makes *some* sense, even if it is not common sense. How could it be that A's speed with respect to B was the same as A's speed with respect to C when C is moving away from B in the same direction as B is moving away from A? Let us say that C is moving eastwards with respect to B at 10 metres per second, and that B is moving eastwards with respect to A at 10 metres per second, then surely C would be moving eastwards with respect to A at 20 metres per second. Surely it could not be said in this case that A had the same speed with respect to B as it had with respect to C? Such a supposition would contradict a basic tenet of Galilean kinematics, that surely everyone accepts, namely that relative velocities are additive.'

Such an assumption would indeed be contrary to Galilean kinematics. And, indeed, the summation of relative velocities seems to work in general. Why, then, deny such a seemingly well-confirmed theory? But it is not so much a matter of wanting to deny Galilean kinematics as wanting to be *able* to deny Galilean kinematics. Let us assume that there were some well-accepted principles and theories of science such that if all but one of them were true, the remaining one would be false. That is, all the theories taken together are contradictory. They cannot all be true. To put it yet another way, it is necessary, by virtue of the logic of any language in which the theories can be expressed, that one of the theories is false. In such a case we shall want to be *able* to deny any of the theories pending investigations as to which of the theories *is* false.

This situation prevails with three widely held theories: Maxwell's theories of electromagnetic radiation, Galilean kinematics, and a principle which we meet now for the first time in this book, the Restricted Principle of Relativity. This principle is that if neither A nor B is undergoing acceleration, then the course of events with respect to A is determined by exactly the same general laws by which the course of events are determined with respect to B, regardless of any relative motion between A and B. Why are these three theories inconsistent with one another?

We saw in the previous section that from Maxwell's equations can be derived the speed of electromagnetic radiation with respect to a medium—or to put it another way, the speed of electromagnetic radiation in a medium with respect to some point that is stationary with respect to the medium. We saw that this gave us a particular velocity with respect to the 'medium' we called empty space. This seemed to give us the conclusion that there was an absolute space with respect to which we could say that some object or thing was stationary. However, the situation turns out to be not quite so simple as this, if we consider the speed of electromagnetic radiation in a medium with respect to a point with respect to which the medium is in motion.

Let us assume, for example, that the medium is moving with a speed u in a certain direction with respect to our reference point, and let us assume further that the speed of electromagnetic radiation with respect to the same reference point in the same direction be v. Let v_s be the speed that the radiation would have were our medium stationary with respect to our reference point. Let ϵ be the permittivity of the medium and ϵ_0 be the permittivity of empty space.

Then it can be shown that Maxwell's laws for electromagnetic radiation entail the following:

$$v = v_s \left\{ 1 - \left(1 - \frac{\epsilon^0}{\epsilon} \right) \frac{2u}{v_s} \right\}^{\frac{1}{2}}$$

Thus, in the case of empty space where $\epsilon = \epsilon_0$,

$$v = v_s = c_0$$

So, even if we can make sense of the assumption that there can be points which are at rest with respect to empty space, we are left with an extraordinary result if we assume the Restricted Principle of Relativity to be true and also accept Maxwell's theory as a law of nature.

For, given the Restricted Principle of Relativity, then if Maxwell's theories *are* laws, they apply equally to a point at rest with respect to space, and to a point moving with unaccelerated motion with respect to space, or, what is the same thing, to a point with respect to which space is an unaccelerated motion. Hence, given our previous result, electromagnetic radiation will have the same speed with respect to *both* points, thus violating the Galilean theory that relative velocities are additive.

If we were to accept this conclusion, then, rather than to reject the premises on which it was based, we would be committed to believing that there is something, namely electromagnetic radiation, which has a relative speed, c_0, with respect to anything else at all, regardless of that thing's relative motion to any third thing. We would have, therefore, a case for describing something as being in absolute motion with speed c_0, in the sense introduced on page 25, which is a sense of 'absolute motion' that does not entail the existence of a Newtonian space.

In section 2.5 we seemed to be left with the alternatives of having to give up either the relationalist programme or Maxwell's theory of electromagnetism. The arguments in this section were aimed at showing that there are still further alternatives. In particular one such alternative is to give up Galilean kinematics. And what this all began with, back in sections 2.2 and 2.3, was an attempt to apply the relationalist programme to some electrical and magnetic properties of space which seemed on the surface of it to be eminently reducible. However, as has been shown, the ramifications of these attempts at reduction are very wide indeed—striking at very fundamental principles in areas of physics that seemed far removed from anything to do with electricity and magnetism. The path which the relationalist must tread is not always an easy one.

2.7 DYNAMIC PROPERTIES OF SPACE

This section deals with an area of physics, namely dynamics, which has been a favourite battleground for dispute between the relationalist and those who believe in an absolute space, since the days of Newton. I shall begin by illustrating the problem for the relationalist with an example which is a variation on a theme by Newton himself. It will become clear, once again, that there is more than one avenue open to the relationalist—but each demands a sacrifice of some cherished belief.

One method that has been suggested for providing artificial 'gravity' for astronauts is the use of a rotating space-ship. Consider such a space-ship as shown in Fig. 8.

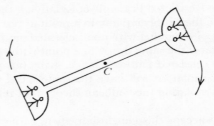

Fig. 8. *A rotating space-ship*

The space-ship consists of two cabins joined by a communication tube. It rotates as shown around its centre of mass, *C*. The floors of the cabins exert forces on the feet of the astronauts, these forces being those required to keep the astronauts rotating about *C*. Without such forces, the astronauts would float off in a straight line away from the space-ship.

Likewise the communications tube must exert forces on the two cabins in order to sustain the rotation. The equal and opposite force-reactions by the cabins on the tube will keep the tube in tension. This tension will produce an increase in length in the communications tube. Thus the astronauts could use a measurement of the increase in length of the tube to determine their rate of rotation without actually looking out of the cabin windows.

Let us assume that the rate of rotation as determined by the astronauts is, say, one revolution per minute. They then look out of the window and measure their rotation with respect to the stars by timing the apparent rotation of the surrounding stars with a stop-watch. Much to their surprise they find that their rate of rotation with respect to the stars is zero. All the stars, then, are rotating about some point or other with a rate of one revolution per minute! A possible story? If not, why not? To put it another way, is it possible for some object to be rotating, and yet not to be rotating with respect to any other massive bodies? Or, again, is it possible that the entire universe is rotating?

'Haven't we been through all this in the previous section?' some reader may ask. The answer is 'not quite'; for here the subject is not just motion, but *change* of motion—acceleration. Further, accelerations of bodies involve forces upon those bodies and these forces are often detectable as stresses within the bodies themselves. The only way to detect unaccelerated motion may be to detect motion with respect to other bodies. With accelerations, however, we do not seem to be so limited.

Assuming for the moment that some accelerations are detectable in a way quite independent of observations of other bodies, we might wonder whether it is possible to detect accelerative forces when there is no acceleration with respect to the remainder of the bodies of the universe. Let us assume further that such a thing was possible, and that such forces had been detected. Then with respect to the frame of reference determined by the 'fixed' stars, Newton's laws of dynamics would not be true, for according to Newton's First and Second Laws of motion, a body is unaccelerated if and only if the sum of the external forces upon the body is zero.

One could, of course, still say that Newton's laws were true with respect to some other frame of reference, which, of course, would have to be one which was accelerating with respect to the fixed stars. But then, of course, one would have to postulate the existence of forces on all the 'fixed' stars, in order to account for their accelerated motion in this reference frame. Let us assume that someone did this. Then one might well ask, 'With respect to what are all these stars accelerating?' For Newton the answer would have been 'space'.

What moves could be made by one who did not wish to countenance such an entity? He might make one of the moves with which we are already familiar. For example, he might wish to deny that acceleration is always acceleration-with-respect-to-something. But then, what he would mean by 'acceleration' would no longer be clear. The new notion of acceleration would need to be explained.

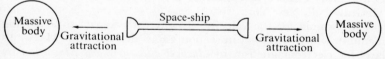

FIG. 9. *Non-centrifugal stress produced in a space-ship by gravitational fields*

Alternatively, one could deny (and very plausibly) that our assumption that the accelerations of some bodies are detectable in a way quite independent of observation of other bodies was true. For example, in the case of the spacecraft mentioned earlier, one might deny that the stresses within the spacecraft were those of centrifugal forces. One could conjecture, for example, that gravitational forces acting in an equal and opposite way on either end of the ship produced the stresses without producing any corresponding acceleration (see Fig. 9). Such an explanation of the stresses would be plausible only on the basis of observations of what objects were in the vicinity outside the space-ship. But, likewise, any explanation of the stresses in terms of centrifugal forces would likewise be plausible only if there were no grounds, on the basis of external observations, for believing that the forces were produced otherwise.

Again, one could deny that there is any frame of reference with respect to which Newtonian dynamics was true. Of course, were one to do this, one would be left with the job of finding some other theory of dynamics which would explain all observed dynamical phenomena, including those of our stressed space-ship—and this would not be easy. The point is that many different reductions of statements about space, arising from the case of the stressed space-ship, are possible. Some of these will be explicatory reductions, some will be theoretical reductions, most will be both, and all will be attempts at an ontological reduction of space.

2.8 SOURCES AND HISTORICAL NOTES FOR CHAPTER 2

Consideration of the electrical and magnetic properties of space as a rationale for absolutism has been decidedly neglected in philosophical circles, though it is clear that such considerations were instrumental in determining the way that Albert Einstein was to interpret Maxwell's electromagnetic theory, and were also instrumental in his accepting that theory so interpreted. See, for example, his autobiography in *Albert Einstein Philosopher—Scientist*, from *the Library of living philosophers*, edited by Paul Arthur Schilpp (Harper and Row, New York, 1959), especially page 35.

There are many useful texts covering the physics described here. One such is Frank's *Introduction to Electricity and Optics* (2nd edn., McGraw-Hill, New York, 1950).

The formula for the velocity of light with respect to a moving medium, which was used in section 2.6, is derived from Maxwell's laws for electromagnetism in *Classical Electricity and Magnetism* by Panofsky and Phillips (the A. W. Series in Advanced Physics, Addison-Wesley, Reading, Mass., 1955).

Further background reading on the theoretical contradiction facing physicists at the beginning of this century, which was outlined in section 2.6, may be found in the autobiography of Albert Einstein mentioned earlier.

Aristotle's views on kinematics may be found in his *Physics*. The Clarendon Press has published translations of all Aristotle's works. The editors are J. A. Smith and W. D. Ross. Ross has provided a commentary entitled *Aristotle's Physics* (Clarendon Press, Oxford, 1936).

Newton's views on kinematics, dynamics, and gravitation were first published in his *Mathematical Principles of Natural Philosophy* in 1687. The *Principles* were first published in Latin. A translation into English by Andrew Motte (1729) has been revised by Florian Cajori and published by the University of California Press, Berkeley, Calif., 1934.

The treatment of the problems of dynamics in section 2.7 is only introductory. The example of the rotating space-ship is a retelling of Newton's example of a pair of globes connected with a cord kept in tension by the rotation of the pair. This example of Newton's occurs in the Scholium to the Definitions of his *Principles* mentioned above. Of all the critics of Newton's theory of an absolute space, perhaps the most famous is Ernst Mach. His views are expressed in Chapter II, section VI, 2–6, of *The Science of Mechanics: A Critical and Historical Account of its Development*, translated by T. J. McCormack (Open Court, La Salle, Ill., 1960).

Einstein's *Relativity—The Special and the General Theory*, Part II (University Paperbacks, Methuen, London, 1960) also contains a discussion of this problem.

Van Fraassen's discussion of the problem in his *An Introduction to the Philosophy of Time and Space* (Random House, New York, 1970), especially Chapter IV, Section 1, is clear, but van Fraassen (page 116) discards the problem of why it is that some reference-frames are Galilean and others are not. Einstein, following Mach and Newton, insists that an explanation *is* called for, though of course he rejects Newton's theories of an absolute space as a satisfactory explanation. Einstein and Mach insist that the inertial properties of any material body should be accountable in terms of the relationships obtaining between that body and other material in the universe. It is this that Einstein and others since have referred to as 'Mach's principle'.

CHAPTER 3

Space and Geometry

3.1 GEOMETRICAL PROPERTIES OF SPACE

One of the assumptions of Newtonian dynamics is that space is
Euclidean, that is, space satisfies the axioms, and hence the theorems,
of three-dimensional Euclidean geometry. Within Einstein's General
Theory of Relativity, on the other hand, it becomes allowable that at
various places space is curved, that is, space is non-Euclidean. What
can this possibly mean?

Two-dimensional Euclidean geometry is the geometry commonly
learned by children at school. We are taught it as the geometry of
the flat plane. The axioms of the geometry as usually taught are five
in number and may be expressed as follows:

1. Given any two points x and y, there is at least one straight line
   on which both x and y lie.
2. Any finite segment of a straight line is part of one and only one
   straight line of infinite length.
3. Given any point x and any distance r, there is one and only one
   circle with centre x and radius r.
4. Any two right angles are of equal magnitude.
5. Given three straight lines, p, q, and r, one of which, p, intersects
   the other two, q and r (see Fig. 10), then if the sum of the

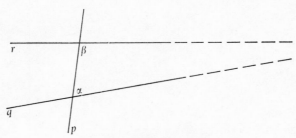

Fig. 10. *Illustrating Euclid's fifth axiom*

interior angles of intersection on the same side of p (e.g. α and β
in Fig. 10) is less than two right angles then, if r and q are
produced indefinitely, they will meet on that side of p.

Strictly speaking, more axioms than these are needed to develop Euclidean geometry, and these have been made explicit by mathematicians from Euclid on including Euclid himself. For example, also needed is the axiom that if three quantities a, b, and c are such that $a = b$ and $b = c$, then $a = c$.

Given these additional axioms, all of which are readily assumed by most people, plus the five mentioned, it can be shown that if q is a line and x is a point which does not lie on q, then there is one and only one straight line through x which is parallel to q, that is, which does not intersect with q.

Before the nineteenth century, mathematicians regarded axioms 1–4 inclusive as 'self-evident' and attempts were made to show that axiom 5 could be proved from the other four plus, of course, the additional axioms mentioned. Euclid had already shown, at least to the satisfaction of his contempories, that it was possible to prove without axiom 5 that, through a point external to a given straight line, there was *at least one* straight-line parallel to the given line. What remained to be proved was that there was *at most* one such line. Many of the attempts to prove this took a *reductio ad absurdum* form. That is, what had to be proved was assumed false and the mathematician then tried to show that a contradiction resulted. In the nineteenth century, Beltrami showed that such a contradiction would be forthcoming if and only if Euclidean geometry was itself inconsistent. Thus a geometry based on the assumption that through a point external to a given line *more than one* parallel could be drawn was shown to be consistent if and only if Euclidean geometry was consistent. Such a geometry called hyperbolic geometry had already been developed in the early nineteenth century by Gauss, Bolyai, and Lobachevsky—all working independently.

Soon after, Riemann had developed spherical geometry—the geometry of the surface of the sphere. Instead of using axiom 5, this geometry employs the assumption that there are no parallel lines, and instead of axiom 2, it used the assumption that any two lines had *two* points in common. This geometry also was shown to be consistent if and only if Euclidean geometry was consistent, as have still other geometries that have been developed since.

Now if different two-dimensional geometries are possible, it seems reasonable to conjecture that different three-dimensional geometries are possible, and, indeed, it can be shown that this is so. The question then arises—which geometry correctly describes the three-dimensional space in which we live? Does space have a Euclidean (or flat) geometry as we have presupposed in the past, or does it have some non-Euclidean (or curved) geometry? Does it make sense to say that space has some one geometry rather than another—and if so, how could we find out which geometry space does have? What evidence would count for or against space exhibiting a particular geometry—Euclidean geometry for example?

One of the consequences of the axioms of Euclidean geometry is the well-known theorem known as Pythagoras' Theorem. It states that, given a right-angled triangle, the square of the distance along the hypotenuse is equal to the sum of the squares of the distances along the other two sides (see Fig. 11).

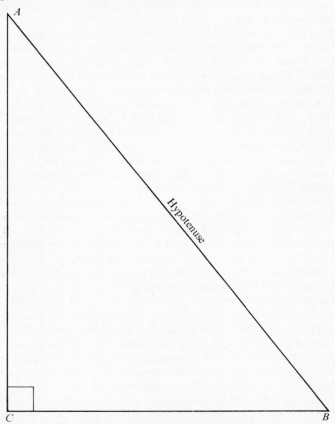

FIG. 11. *Given that* ABC *is a triangle with angle* ACB *equal to one right angle, then it follows from the axioms of Euclidean geometry that* $AB^2 = BC^2 + CA^2$

Therefore, it would appear that we could set up an experiment to test whether or not space has a Euclidean geometry in the following way. We mark out three points *A*, *B*, and *C* such that the angle

$A C B$ is a right angle. We then proceed to measure the distance along a straight line from A to B and likewise from B to C and from C to A. Call the distance from A to B, AB; the distance from B to C, BC; the distance from C to A, CA. We then check to see if $AB^2 = BC^2 + CA^2$.

If this result was not realized within the limits of the accuracy of our measurements, we could say that space did not exemplify a Euclidean geometry—or could we?

The first difficulty we would meet in trying to carry out the above experiment occurs with the instruction 'We mark out three points A, B, and C such that . . .' What is it to mark out a point of space? Points of space, unlike birds, fish, or motor-cars, are not material objects that can be tagged. Indeed, just as one may have ontological worries about space, one may also have ontological worries about points of space. For, given the existence of spatial points, we could regard space as the aggregate of all such points. This remark is particularly relevant to the discussion earlier on the kinematic properties of space (section 2.6, pages 23 to 27) especially the discussion involving Maxwell's laws. There remarks were made concerning points of reference and the belief that electromagnetic radiation has the same velocity with respect to two different points moving with respect to one another. And this raises another difficulty. Our points A, B, and C had better not move with respect to one another or the results of our experiments will not be consistent. But this is at most a practical difficulty. In principle we could come as close as we liked to 'freezing' our spatial points by carrying out the experiment in an appropriately short time . 'Moving' points would then coincide, within the limits of experimental error with points which were at 'rest'.

But let us ignore both these difficulties for the moment and assume that three mountain-tops have been chosen for our points A, B, and C. That is, the points A, B, and C are tagged by the tops of the mountains. BC is measured to be exactly three kilometres, CA is measured to be exactly four kilometres and hence AB should turn out to be exactly five kilometres ($3^2 + 4^2 = 5^2$). However, when measured, AB is found to be 4·5 kilometres.

Would a physicist who heard of such a result immediately think that Euclidean geometry had been shown to be false? There are at least two things that he would want to know. Firstly, were the distances BC, CA, and AB measured along straight lines, and how was the straightness of these lines determined? Secondly, how were the distances determined? Of course, the experiment would be rejected out of hand had it been known that the distances were measured by rolling a wheel of known circumference from A to B,

B to *C*, and *C* to *A*, up hill and down dale, making sure to divert around Farmer Brown's pig pen and Farmer Green's wheat field. The distances that the wheels rolled may not have been queried so much as the fact that they had not been rolled in straight lines. But what then should the experimenter use to determine a straight line? Should he use the limiting shape of a tightly stretched string, or should he use line of sight (that is, the paths of light rays). Or again should he assume in his measurements, that the shortest distance between two points is a straight line? All these are techniques which are commonly used for determining straight lines. Why should we believe that any of them actually do so?

It follows from Newton's laws of force that the limiting shape of a stretched string should be a straight line—otherwise there would be unbalanced forces on some section of the stretched string which would cause it to move laterally into another position. Again it follows from Maxwell's laws that given a homogeneous medium, electromagnetic radiation (and hence light), moves in straight lines. Finally, it is axiomatic in Euclidean geometry itself that the shortest distance between two points determines a straight line.

Doubtless our experimenter could have used other techniques also for determining his 'straight' lines; but let us examine these three. Firstly, the stretched string technique. We have said that it follows from Newton's laws that the limiting shape of a stretched string is a straight line. A physicist who knew that the experimenter had used this technique in achieving his non-Euclidean result, would therefore have an alternative hypothesis to explain the anomaly. Rather than denying Euclidean geometry, he could wonder whether Newton's laws were false. In practice, of course, he would probably doubt the veracity of the experimenter's results, but the point here is that, even if he did not doubt the experimental results, even if he thought the string was being held sufficiently taut, he would not have to conclude that Euclidean geometry was false.

Similarly with the light-ray technique. Even if the physicist were sure that the atmosphere between the mountain-tops was homogeneous and so on, he could always wonder whether Maxwell's laws of electromagnetic radiation were correct rather than deny Euclidean geometry.

Poincaré claimed that physicists would always modify the laws of optics rather than reject Euclidean geometry. Further, he claimed that a belief in Euclidean geometry was consistent with the results (or strictly with the statement of the results) of any experiments or observations.

In the former claim he was wrong. As a result of an observation made by an expedition organised by Eddington and Cottingham in

1919, involving the passage of light rays originating from distant stars and passing close to the sun, physicists tended to reject Euclidean geometry rather than Maxwell's laws. But how about the second claim? Is it always possible to reject some other theory and continue to accept Euclidean geometry no matter what the experimental results may be?

How about the third method of determining straight lines? This doesn't seem to depend on any physical theory outside of Euclidean geometry. But the *distances* still have to be measured—and there are many methods at hand for measuring distances.

Let us examine just one method which could be used—the method which in the past has been considered to be 'fundamental' to many physicists and philosophers of physics. The method is to acquire some rigid rods. Make sure that they are of the same length by placing them one on top of the other and observing that their ends coincide. Then place them end to end along the distance to be measured. Counting the number of rods needed to traverse the distance would give us a measure of the distance. The shortest such measure between *A* and *B* would then be along a straight-line path between *A* and *B*.

Of course, there is also the business of determining that angle *ACB* is a right angle, but once again this could be done by using manipulation of rigid rods in a method that is familiar to most secondary-school pupils. Now let us assume that with such measurements carefully made, our experimenter achieves the result previously mentioned, namely that the ratio of *AB* to *BC* to *CA* is as 4·5 is to 3 is to 4.

How can our physicist fail to reject Euclidean geometry now? One way for him to do this is to postulate that, although the rods all have the same length when placed upon one another, they *change* their lengths when moved about from one place to another. This hypothesis could, in turn, invite other physicists to *explain* this postulated change in length. In practice, this is common enough. Change in lengths of otherwise rigid rods is often explained in terms of changes in temperature, changes in forces acting on the rods, and so on. But, if such causes as these were operative in changing the lengths of the rods as our experimenter transported them about, then, since different materials have different elasticities and different coefficients of expansion with temperature, we should expect different results with different materials. Let us say our experimenter now uses rods made of a different material and let us say that to all intents and purposes the same results are obtained. He repeats the experiment again and again each time using different rod materials on different occasions. Still he obtains the same result. Let us say that, as a result of such experiments, physicists conclude that, if there is a change in length of the rods, the change has nothing to do with the material of which the rods are made.

Let us now review the options open to our physicists. They could assume:

(a) that space has a non-Euclidean geometry;

(b) that space has a Euclidean geometry, but that different areas of space had different effects upon the dimensions of bodies;

(c) that Euclidean geometry is everywhere applicable, but that space is permeated throughout by something or other (a field, or an aether) variations within which affected the dimensions of all bodies in a similar way as they are transported about;

(d) one could forget about space altogether and simply claim that the method of measuring distances by the superposition and transport of rigid rods does not in general yield Euclidean results.

Many a relationalist would not want to go further than this latter claim. They would argue that all talk of space having this rather than that geometry should either be eliminated or reduced in favour of statements concerning the geometry of systems for measuring distances and for determining straight lines. Since what is a straight line and what distances are deemed to be equal will depend on the measuring system, very many geometries are likely to be applicable.

However, it might be objected that this approach would make physics enormously complicated. For since measuring systems for determining distances and straight lines can be infinitely varied, and since the notions of straight line and distance are needed in kinematics, dynamics, and electromagnetics, we will not now have one physics—but an infinity of them.

Further, if we are going to regard one measuring system for distances and the determination of straight lines as being equally good as any other, why not adopt the same approach for forces, durations of time, and charges and masses, and thereby have our dynamics and electromagnetic theory a function of different measurement systems also, systems not only for the measurement of length, but of forces, times, charges, and masses as well. That is, why not place Newtonian dynamics, Maxwell's electromagnetic theory, and Euclidean geometry all on the same footing? Thus one set of measurement systems will be, say, Euclidean, Newtonian, and Maxwellian. Another may be none of these. A third may be Euclidean, but not Newtonian or Maxwellian.

Some physicists, for example P. W. Bridgman, would be content with this approach to physics. Indeed they would claim that would-be statements of physics are meaningless unless the parameters involved in the statements are considered to be determined by some particular set of operations or measurements. Thus the so-called laws of physics such as Newton's laws, the theory that space has a Euclidean geometry, Van der Waals's equation, and so on are not simply true or false according to these physicists, but rather true *of* or false *of* particular sets of measurement operations.

But most physicists have not adopted this approach, for a number of reasons. The first is simply repugnance at the manner in which this approach, if taken seriously, would complicate the subject. As previously stated, instead of there being simply one physics there would be an infinity of physics—one for each set of measurement operations.

One way out of this would be for physicists to agree on certain particular measurement operations and base their physics upon these alone. But this would have the disadvantage that it would seem to leave quite a lot of physics unsaid. What 'laws of nature' would other conventions yield? This might not be such a problem if the conventions adopted were everywhere and in all circumstances applicable, for then one could study and measure the sort of operations used in the other systems with the conventions adopted, and, if one were lucky, one might be able to calculate what laws were applicable to other measurement systems. The trouble is that there are no measurement systems, for length, mass, time, charge, force, or any other physical quantity, which are everywhere and in all circumstances applicable. For example, we cannot use the superposition and transportation of rigid rods for astronomical distance measurements, and likewise we cannot use observations made on cepheid variable stars for distance measurements in the laboratory. Thus this approach would mean that we could not deal with the distances of distant galaxies if we could measure distances between objects in the laboratory and, if we could deal with distances in the laboratory, we would be frustrated with respect to astronomical distances—unless of course we adopted one measurement convention for each of the two kinds of 'distance'. But then we could not assert such facts as that Andromeda is much further away from me than this table for there would be no guarantee that the 'distance' of Andromeda was in any way related to the 'distance' of the table, the two 'distances' being determined by completely different operations.

An alternative approach would be not to convene on measurement systems at all, but rather to restrict them by imposing upon them certain conditions. For example, we might insist that measuring operations for distances must not violate Euclidean geometry, while measuring operations for temporal durations, forces, and masses must not violate Newton's dynamics and Galilean kinematics. Of course, in doing this we would be making Euclidean geometry and Newtonian dynamics unfalsifiable by experiment and some, for example Karl Popper, would claim that they had thereby become pseudo-scientific statements only. Is it the case, however, that all empirical content would be lost from such theories if we adopted them as standards which measurement systems had to satisfy? Conceivably, not all empirical content would be lost. For it would still be a factual matter whether or not there were measurement systems which satisfied these standards. For example, if physicists began to despair of finding a set of measurement systems for forces, masses, distances, and times which yielded results which were always consistent with, say, Newtonian dynamics, then they could well consider adopting a new standard, that is a new criterion for the correctness of measurement systems, namely a different system of dynamics. It would be at such a point that physics could experience what Kuhn has called a 'scientific revolution'.

Is this, then, the way physics is? Not quite. It is true, of course, that physicists tend to be fairly conservative with respect to 'well-established' generalizations. Thus we can see why, as beliefs become 'well-established', they are likely to become even better-established. More and more will be given up in their favour—including systems of measurement. But life for the physicist is not quite as simple as this. Occasionally, well-entrenched theories are found to be mutually inconsistent as was shown in section 2.6. Further, physicists can be conservatively-minded about matters other than their favourite theories. If it came to a choice between the superposition and transportation of rigid rods as an allowable distance-measuring system, and Euclidean geometry as a standard for measuring systems, my own guess is that there would be about as many choosing one way as the other. Further, there are the everyday physical facts of life with which all theories and measurement systems must square— for example, that this cup is considerably smaller than this house; that Sydney is considerably further from London than Paris is from Rotterdam. Indeed the physicist, as a rule, is hard put to find a system of beliefs and measurements which will be self-consistent and at the same time satisfy most of the conservative pressures upon him. In practice, he is far removed from having the wealth of free choice which we envisaged earlier.

But in spite of all this interesting sociology, and in spite of the plausibility of the reduction of statements about the geometry of space to statements about measuring systems, the question still arises as to what is being measured with these measurement systems. By virtue of what do the measurements yield the values they do?

These problems will be discussed in section 3.6. Meanwhile, let us examine a couple of properties closely related to the geometrical properties we have just been discussing.

3.2 TOPOLOGICAL PROPERTIES OF SPACE

If what was said in the previous section is correct it would seem that we could reduce talk of the geometric properties of space to talk of the geometric properties of measuring systems—either actual or prospective. The absolutist could always object, of course, that though the geometrical results of our measurements were explained by his theories of an absolute space wherein lay all our measuring instruments, the relationalist could not explain his results in this way.

The relationalist could, of course, reply that the absolutist is taking his desire for explanation too far—that just as theists have invented a god to 'explain' why the laws of nature are what they are, so the absolutist has invented a space to explain why bodies obey the geometrical laws that they obey. In both cases the relationalist could well invoke Occam's Razor, the principle by which philosophers are asked not to multiply entities beyond necessity.

To support the relationalist's position, Poincaré argued that everything we observed was a material body and that since no amount of observation was observation of empty space—all that we could learn from observation was the relationships of material bodies and nothing about the properties of space at all. Analogously, he argued, no amount of observation of a ship's timbers could bear on the age of the captain. But this analogy will not do. If a measurement of the ship's timbers revealed that the ship was 100 metres long by 20 metres wide, and we had good reason to believe that all ships approximating to these dimensions were ships of the Ruritanian Navy and further that all Ruritanian Navy sea-captains were obliged to retire at forty years of age, then measurements of a ship's timbers *could* have an evidential bearing on the age of the ship's captain. The point is that any contingent proposition, p, can be evidence for any other contingent proposition, q, provided that p and q are consistent with one another and provided also that there is good reason to believe the truth of some further proposition or conjunction of propositions, r, such that p and r together are regarded as good evidence for q. The possible existence of such a proposition r is guaranteed by the fact that the corresponding statement of the form 'if p then q' would always do the job, provided there was independent evidence for the truth of this proposition. So Poincaré's analogy is irrelevant. Nevertheless it may be correct that if p is a statement about space, and q is a statement about events, then there can be no independent evidence for the statement 'if p then q'. If so, Poincaré would be correct in believing that statements about space are epistemologically isolated. More will be said of arguments of this sort in section 3.4. Meanwhile let us assume that experiments with measuring instruments do not give the relationalist any worries. Are there any observations, that do not involve measurements, which might do so?

Imagine a prison-cell. The walls are impervious—likewise the roof and the floor. Prisoners enter through a door in the wall which is then securely barred behind them. Nevertheless, much to the consternation of the legal fraternity, the prisoners keep appearing on the outside of the walls. It is found that they do not escape through the walls, neither do they tunnel out through the floor, nor do they break out through the roof. When queried, all that our band of escaped prisoners can tell us is that they walked towards the *centre* of the prison-cell, kept walking in that direction, and ended up on the outside of the walls.

Some reader may think that this description of a prison, its prisoners, and ex-prisoners is inconsistent, but they would be mistaken. For everyone is perfectly familiar with the two-dimensional analogue of this situation. Consider the two-dimensional surface of the doughnut-like figure of Fig. 12. Around the hole in the doughnut we draw a one-dimensional 'prison-wall', *W*. Any prisoner on the

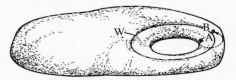

FIG. 12. *A two-dimensional toroidal universe with a badly sited prison wall* W

two-dimensional space who was within *W* at *A*, say, could move towards the centre of *W* and, continuing in the same direction, end up at *B* on the outside of *W*, without, of course, passing through *W*. To put it another way, toroids have the property that it is possible to draw closed curves on their surfaces, without dividing the surface into two distinct portions. On the other hand, the surfaces of spheres, for example, do not have this property. Any 'impenetrable', closed, one-dimensional curve on the surface of a sphere is an effective 'prison' for any two-dimensional being living on the sphere. For these reasons it is said that the *topology* of a sphere is different from that of a toroid.

Now if we were to find that bodies could move from the inside of some closed two-dimensional surface to the outside or vice versa, without passing through the surface, would not this tell us something of the topology of the three-dimensional space in which we live?

A relationalist, of course, would still wish to deny this. At the most, he would say, it would tell us that bodies can move *as if* they moved on a three-dimensional toroidal surface. Could one reasonably regard it as analytic that the study of the geometry and topology of physical space was the study of the *ways* in which it was physically possible for bodies to move or position themselves in relationship to one another? But since the same question arises with respect to the dimensionality of space, let's talk about that first.

3.3 THE DIMENSIONALITY OF SPACE

Nobody, so far as I know, has ever built a prison-cell from which prisoners can escape without going out through a hole made in the walls or roof or floor, and most jailers, I guess, would be fairly sure in the belief that any escapees will not be achieving their freedom by anything but the standard methods of breaking out through the prison surface. The confidence that jailers have in their cells will therefore be approximately proportional to the extent to which they believe that the walls, roof, and floor are impervious.

However, consider the following situation. A policeman sees two burglars escaping from a building. He quickly apprehends one of them, but, in order to leave himself free to chase the other, he decides to imprison the man he has caught by drawing a chalk-mark around him on the pavement. The prisoner then removes his handkerchief from his pocket, rubs out some of the chalk, and makes his escape by walking out through the gap thus made. The point of this comedy, of course, is that the prisoner did not have to make a hole in the 'wall' of his prison in order to escape. He simply had to step over it.

What is more, we rather think that every one-dimensional 'prison wall' is like this in our world. That is, we believe that any body is able to move in a *third* dimension away from any two-dimensional surface in which lies a closed one-dimensional curve encircling the body. This is why, then, we believe that space has at least three dimensions. If we believed that it had more than three dimensions we would believe likewise that imprisonment within a two-dimensional surface was impossible. But alas, we do not and for good reason, and so we believe that space has three and only three dimensions.

It would seem reasonable, then, to suggest that statements asserting that space is three-dimensional could be reduced to statements asserting that the ways in which it is physically possible for bodies to move and position themselves with respect to one another are three-dimensional. A similar reduction was suggested at the end of the last section for statements about the topological properties of space. In the following section an attempt is made to generalize this kind of reduction for all cases in which space is said to possess some property universally.

3.4 SPACE AS PHYSICALLY POSSIBLE TYPES OF EVENTS

In the last two sections it was suggested that certain statements about space might be reducible to statements about ways in which it is physically possible for bodies to move or position themselves. These statements were both statements in which space was alleged to possess universally some particular property. The question arises as to whether or not a similar reduction, in terms of physically possible events, is always possible for true contingent statements which assert of space that it possesses some ubiquitous property, Ψ.

In the opening paragraphs of section 3.2, it was mentioned that Poincaré had claimed that such reductions are always possible. In this section I shall endeavour to develop Poincaré's argument in detail, without recourse to the analogy about the ship and the captain which was criticized in section 3.2.

Poincaré's basic premisses are that any observations we make are the effects of material causes, and that space is not matter. Consistent with these assumptions, which seem plausible enough, two cases arise for the statement that space is Ψ. In the first case, space being Ψ could not be causally associated with any properties or distribution of matter, and in the second case space being Ψ could be so associated. In the first case there would be no possibility of anyone knowing that space is Ψ. Someone might believe that it was the case but that belief would be completely causally dissociated from the fact that space was Ψ. Now there is, of course, no need to bother the relationalist with statements which as far as we know are as like as not false. There is pressure to believe in Santa Claus only in so far as some statement about Santa Claus is known to be true. If there is no knowledge that any such statement is true, then an ontological reduction of contexts involving 'Santa Claus' seems in order.

Let us turn, then, to the case when space being Ψ does causally affect the properties and distribution of matter. The effect then of space being Ψ will be to limit the types of physically possible events involving matter to, say, events of type E. Since by our hypothesis space is ubiquitously Ψ, then all events will be of type E. But if space were always Ψ and events involving matter were always of type E, then how could we know that space being Ψ *caused* all events involving matter to be of type E? There would be no possibility of varying the Ψ-ness of space in order to observe whether or not events involving matter and being of a type different to that of type E occurred. Similarly, neither could there be any evidence for the inverse causal relation. So *unlike* the case of the ship's timbers and the captain's age, there is no possibility here of finding evidence for some 'if . . . then . . .' statement which would link 'space is Ψ' to 'all events involving matter are of type E'. For any evidence we would have that 'space is Ψ' would be via a belief in a theory to the effect that such an 'if . . . then . . .' statement was correct. But if the only evidence available for this 'if . . . then . . .' proposition involved our belief in the Ψ-ness of space, the whole exercise would be circular. Under these circumstances, it would appear eminently reasonable for the relationalist to reduce the statement that space is Ψ to the statement that all events are of type E, the reduction being onto-logical. All the explanatory power of 'space is Ψ' would be carried by 'All events are of type E' with the exception, of course, of the explanation of why it is that all events are of type E.

It is this exception that is often the parting of the ways between those who seek a relationalist reduction and those who do not. Those who do not want the reduction will be those who feel, in a particular case, that the statement that all events are of type E cries out for explanation and that that explanation is provided by the statement that space is Ψ. The need for explanation in some cases is obscured here by our use of the words 'events of type E', as if 'E' were some very simple description. But 'E' may be a disjunction of very complex descriptions. Take, for example, once again, the case of the state-ment 'space has a permittivity of 8.55×10^{-12} farads per metre.' In section 2.5, the relationalist was suggesting that this statement be reduced in terms of events which were *either* events involving the behaviour of electric charges on the plates of empty capacitors *or* events involving the behaviour of electromagnetic radiation when no medium was present. But these disjuncts, on the surface of it, present very different descriptions of events. The theory that space has a permittivity of 8.55×10^{-12} farads per metre had the explana-tory advantage of relating these different descriptions. If the relationalist's reduction is not to lose this explanatory advantage, the theory that space has a permittivity of 8.55×10^{-12} farads per metre must be replaced by a theory of events such that both the behaviour of charges on empty capacitors, and the behaviour of electromagnetic radiation where there is no medium, are both explained by the theory. In this particular case, of course, the theory is available. Maxwell's theory of electromagnetism is powerful enough to cover both sorts of events. It is when such a theory is not

readily available that the sense of loss in the relationalist reduction is
felt most strongly.

3.5   POINTS AND POINTING SYSTEMS

In section 3.1, it was suggested that statements concerning the
geometry of space could be reduced to statements concerning either
actual or possible measurement systems.

However, any geometer will also make reference to such things as
points, lines, and planes. What is to be made of these entities? If one
has not provided a reduction for statements concerning spatial
points, one has not succeeded in providing a reduction for statements
concerning space, for what is surely the case is that the aggregate of
all spatial points *is* space.

It would be tempting to think of spatial points as a mathematical
fiction invented by geometers for their own peculiar purposes. After
all, when I believe that I left my bicycle at a point half-way between
the public house and the post office, I do not expect to find two
things half-way between the public house and the post office—my
bicycle and a point which is coincident with it. Points are not substantial. Surely there is nothing physical that has zero size in any
dimension let alone in every dimension. Nothing physical? Well,
nothing substantial anyway. For points, lines, and two-dimensional
surfaces can be just as important to the physicist as to pure mathematicians. Surfaces of bodies as well as surfaces of equal pressure,
temperature, electrostatic field strength, and so on—are all physical
even though they are only two-dimensional. The intersections of
such surfaces are one-dimensional lines—often of great practical
significance, for example contour lines for the hill farmer, and the
intersection of the water table and the surface of the ore body for
the mining engineer. Physical lines can likewise intersect in physical
points.

No one seems to be greatly perturbed that such points, lines, and
surfaces as these exist—not even those who claim that their ontology
does not extend beyond spatially three-dimensional, time-enduring,
substantial objects—for the existence of these points, lines, and
surfaces seems so obviously subservient to the existence of three-
dimensional, time-enduring, substantial objects.

But unlike the imaginary points, lines, and surfaces of abstract
geometry, these physical points, lines, and surfaces are very badly
behaved. They move about with respect to one another. Even if, by
any measurement criterion, the distances between all these points
could be known at any one instant, the relative positions of these
points would change beyond recognition almost instantaneously.
Insubstantial points, unlike substantial bodies, can move at any

speed—consistent with any theory of dynamics. For example the intersection of two almost parallel wavefronts moves at a speed inversely proportional to the angle between the two points. Indeed, these points cannot be regarded as the points of spatial geometry as we ordinarily know it, for the distances between geometrical points are fixed. Let us say that a physical surface is a boundary determined by a physical property of a substance. Physical lines and points, then, are the intersections of these boundaries.

It may be objected that though our geometrically derogatory comments apply to many physical points, lines, and surfaces, there are nevertheless sets of physical points, lines, and surfaces throughout space, which do, in fact, remain more or less fixed in their distances from one another. The aggregate of such a set of points, lines, and surfaces would, of course, constitute space. Any one of these sets of points, lines, and surfaces could be used as the basis of a measuring system.

Of course there could be an infinity of such sets of points depending on our choice of reference bodies and measurement systems. Which, if any, of the systems, would yield the 'correct' set of points? There are a number of presuppositions to this question which need to be made clear.

The first is that there is a set of physical points, lines, and surfaces such that each point retains its individual identity throughout all time.

The second is that, of any two points A and B in the set, the distance from A to B always remains constant.

The third is that there is only one such set with respect to which physical occurrences can be properly described.

The Principles of Relativity imply the falsity of the third presupposition. These principles would imply that if there were any such sets of points as described in the first and second presuppositions then there are an infinity of such sets.

The second presupposition is one which has received a lot of attention in the literature in recent years, especially at the hands of Adolf Grünbaum and his critics. Grünbaum, as I understand him, would not so much be concerned to say that the distance from A to B sometimes varies, but rather that there is no such thing as *the* distance from A to B. For Grünbaum, a distance is not just a function of two points, but rather of two points and what he calls a 'congruence convention'. This term will be explained in the next section. The proposition that distance is a function not only of the two points concerned but of something conventional as well, is, I take it, at least part of what Grünbaum means when he says that space is metrically amorphous or that space has no intrinsic metric. A metric, by the way, is just a name for whatever function it is that determines a distance given a pair of points. A measurement system worthy of the name will determine a particular metric. For the mathematician working with an abstract set of points, the metric will be a formula. To say that space has no intrinsic metric, then, is to say that any function from pairs of points to distances is not itself dependent solely upon space or its properties.

Both the second and third presuppositions entail or at least presuppose the first; yet the truth of this presupposition is not guaranteed *a priori*.

It may turn out that there is no set of physical points, lines, and surfaces such that each point, line, or surface retains its individual identity throughout all time. And even if there were, there would be no guarantee that such a set would be universally useful as a basis for a spatial measuring system. Consistent with the existence of such a set there could be volumes of space within which there were no physical boundaries at all; where the properties of whatever substances lay within the volume were constant throughout the entire volume. Indeed the volume might contain nothing substantial whatsoever. Let us say that there is such a volume and that it is spherical in shape. Would we be able to refer to any points within such a sphere, for example its centre? Would we be able to carry out any spatial measurement within the sphere? Would it make any sense to say that there was a point at the centre of the sphere which was, say, three metres from the surface of the sphere? The answer to the first question is 'Yes we could; but consistent with our assumptions, such points could not be physical points in the sense that the term "physical points" has been used here.' The answer to the second question is 'Yes we could; but not of course using any pre-existing surfaces, lines, or points as the basis for our measuring system. Our measuring system would have to contain provision for the construction of such surfaces, lines, and points, in cases where they were absent for example, by the introduction of rays of light or perhaps by the introduction of rigid rods laid end to end. Needless to say, however, once this was done, the volume would no longer be empty or otherwise homogeneous.' The answer to the third question must be 'Yes' if the answers to the first and second questions are 'Yes'. But if the points in the empty sphere are not physical points, what are they? Given our assumptions, the only solution would seem to be that their existence would be hypothetical. That is, they *would be* the intersections of certain surfaces we could introduce into the sphere, *if* we *were* to introduce those surfaces into the sphere, and the metrical relations of these points would be the metrical relations that these intersections *would have*, *were* they to be constructed.

The same considerations would seem to apply even if the sphere was filled with a homogeneous substance—including a substantial space like Newton's—*unless* it could be shown how such a substantial space could carry with it its own surfaces, lines, and points which were quite independent of possible applications of possible measuring systems. Incidentally, it would not be sufficient to indicate a possible mathematical model for such a space. For if the proposed surfaces, lines, and points were not to fall foul of the relationalist's tendencies towards ontological reduction, the model would have to be related to physical theory in a way that was epistemically relevant. What must be done, no matter how meaningful or true statements about such points may be, is to show how it would be possible for the existence of these entities to affect our knowledge of the world.

In passing, it must be noted that the attitude of a relationalist who adopted this approach would not be that of a verificationist, that is one who thought that a statement has no physical significance, or is not worthy of consideration by a physicist, if there is no way of knowing whether or not the statement is true. The relationalist applying an ontological reduction, could simply be someone who refuses to believe in the existence of entities given that, at the time, he has no good reason to believe in them. This is not to say that he insists on believing that such entities do not exist, nor does it mean that he necessarily decries the work of those who are endeavouring to find out or invent possible ways of coming to know of the existence of these entities. On the contrary, given that he is a physicist worthy of the name, he will always encourage such efforts.

3.6 AMOUNTS OF SPACE

When the Anglo-Saxon farmer ploughed a furrow, the plough-share passed through a furlong of soil. What could be more reasonable than to say that the strip of upturned earth had a property, namely, that it had a length of one furlong?

When I go to London, the journey is about fifty miles. But by virtue of what is the journey this length? Well, I pass through about fifty miles of English countryside, do I not? When we say that the journey from here to London is about fifty miles, do we mean by this that the amount of ground that would be covered on the journey from here to London would be about fifty miles of ground? It seems plausible.

Now when an astronaut flies in a rocket to the moon he travels 238,000 miles. But what has he passed through 238,000 miles of? Space? Is the 238,000 miles, then, a measure of an amount of space? If there is an amount of space between here and the moon, does it follow that space is substantial—like the English countryside and the soil in the furrow?

Descartes's argument to the conclusion that all supposedly empty space was substantial was based on similar considerations. He claimed, plausibly enough, that it is contradictory that there should be extension that is the extension of nothing, whence he concluded that since there is extension in supposedly empty space there must be substance there as well.

But the conclusion does not follow. For an extra premiss is needed —namely that everything is substantial. But as we saw in the last section this is not so. It is, as we have seen, often possible to produce reductions of statements which apparently make reference to things such that the apparent reference disappears in the reduction. That Descartes's hidden premise is false is clear from the fact that we do not regard numbers, points, surfaces, and shapes as substantial objects—but they are all things to which we can make reference. Even if it were true that making reference to some or all of these things was systematically misleading, it would not follow that such a reference would imply the existence of something substantial. Measurements of mass are measurements of an amount of matter. Length, area, or even volume measurements are not. Such facts were known even to wily Egyptian traders in the days of the pharaohs.

The following question still arises, however: by virtue of what is there just that much space between the earth and the moon, or, on a more domestic plane, by virtue of what is there more space in the living-room than in the bathroom? Certainly it is not by virtue of any substance or substances to be found in the living-room.

Some possible answers to this question have already been investigated in section 3.1. It was suggested there that the causes and reasons for our considering one distance greater than another were a very complicated business indeed, involving well-entrenched theories, well-established facts, well-tried measurement systems, and the relationship between these.

'But', someone may insist, 'this does not answer the question of what it is that bears the property of being so many feet or miles long, it only tells us something of the sociology involved with beliefs concerning distances'.

The second part of this remark is true. What was said towards the end of section 3.1 was a bit of sociology. The first part of the remark, however, is confused. The question under discussion was what is it by virtue of which there are 238,000 miles of space between the earth and the moon? This is not the same question as 'What is it that has the property of being 238,000 miles long?' The answer to that is simple. It is the space between the earth and the moon that is

238,000 miles long, and, for the sake of any relationalist who might be perturbed at our mentioning space in this way, we could go on to point out that such a statement may be reduced to the statement that the earth and the moon are 238,000 miles apart. If the objector now wishes to ask what things bear the relation of being 238,000 miles apart, the answer to that is simple also. It is 'the earth and the moon'. But once again, that was not what was being discussed. The point is that there is a difference between asking what has such and such a property, or what thing bears such and such a relation to what, and asking another entirely different question, namely, 'On what things does the existence of instantiations of these properties and relations depend?' Let me illustrate with an example. Tom is the cousin of Luke. Now if we ask between what things does the relation 'is the cousin of' obtain, the answer is 'Tom and Luke'. But if we ask on what things does the existence of this relationship depend, the answer will be a list including, besides Tom and Luke, a parent of each of them and at least one person who is a grandparent of both of them, plus certain actions and events each involving at least one of these people. To give this list and appropriately to describe the people, things, and events in the list is to explain what the relationship of cousinhood amounts to.

Now back to the sociological remarks at the end of section 3.1. What they tell us is what people look for and what they take into account when they are trying to reach or justify beliefs about distances. Finding out what people would do, say, or take into account, in the justification of their belief in some proposition that they have expressed, often provides useful clues to what they mean by the sentence which they used to express the proposition. This is *not* to say, by the way, that one needs to know what methods of justification someone would use before one knows what that person means.

However, in this section, I shall continue to investigate this sociology with a view to finding out what sort of measurements a physicist would count as a measurement of length. I shall then argue that the restrictions the physicist places on length measurements are such that it is possible that these could yield a *unique* metric. There will be no need in the development of this conclusion to invoke an absolute space as that thing by virtue of which the distances are what they are. Hence it will be claimed that it is consistent to believe in absolute distances and lengths without having to believe in absolute space.

One thing that seems clear from this sociology is that distances are deemed to satisfy a number of propositions. For example: given any four points A, B, C, and D, the distance AB is either less than, greater than, or equal to the distance CD. Further, if AB is greater than CD and CD is greater than EF, then AB is greater than EF. Again, the distance from any point A to itself is always zero. The distance from A to B is the same as the distance from B to A. The distance from A to C via B is the sum of the distances from A to B and from B to C. The distance from A to B plus the distance from B to C is always greater than or equal to the (shortest) distance from A to C. Actually, by distance in this paragraph, I have meant shortest distance all along. But by *the* distance, we usually do mean the shortest distance unless otherwise specified. Another supposition about distances: if the distance between A and B is zero, then A and B are the same point.

Any distance-measuring system worthy of the name would have to satisfy these criteria. Let us call these criteria, plus any others that one may deem essential for things to be called distances, the axioms for distance. The question arises: is there any measuring system that satisfies the axioms of distance? Is there one and only one? Are there more than one?

Some might say that, if the axioms of distance included only the axioms so far mentioned, then either no measuring system would satisfy the axioms—or an infinity of measuring systems would do so. To put it another way, these people might claim that if there is one way of imposing a metric on physical space, then there is an infinity of such ways. Why would someone make such a claim?

Consider a circular table. A physicist is using light rays and the manipulation of rigid rods to determine straight lines, angles, and distances. After a great deal of intensive observation, he concludes that Pythagoras' Theorem is obeyed throughout the surface of the table and that therefore the surface is a Euclidean plane. Now in this process the physicist always took his rigid rod as measuring out one unit, say, of length. What would happen if he took his rigid rod as measuring out one unit of length if it was oriented along a radius from the centre of the table, and as measuring out a distance of

$1 - \dfrac{r \sin \theta}{2(1 - r)}$ if the rod is oriented otherwise (where r is the distance

of the centre of the rod from the centre of the table, and θ is the angle between the orientation of the rod and the radius joining the centre of the table to the centre of the rod). With respect to this 'measurement system' the surface of the table would no longer be Euclidean. It would have some other geometry. We could, of course, by different formulations, generate an infinity of different metrics for the surface of our table in this way. Some of these would yield Euclidean geometries as our first metric did, others would yield different geometries.

Grünbaum would say that these different metrics resulted from different congruence conventions. Now, while it is true that there are an infinity of congruence conventions if there is at least one, the question is: 'Do all congruence conventions count as measurement systems?'

I would claim that not all would, for the following reasons. If a measurement system of physical distances is worthy of the name, it should be one such that it is not necessary for anyone using the system to know either the position of the points between which he is measuring the distance, with respect to any third point, or the orientation of the line joining the two points with respect to any other line or lines, or anything which would entail this information. If we accept this restriction on our measurement system, then the system whereby the physicist counted the length of his rod as one unit, regardless of where it was and regardless of its orientation, was indeed a measurement system. But the congruence convention whereby he counted the length of his rod as $1 - \dfrac{r \sin \theta}{2(1 - r)}$ was not a measurement system for he would have had to know the orientation and position of the rod with respect to the *centre of the table* before he could have applied the system.

'Wait a while,' someone may object, 'sometimes it is very important to adjust the results of our measurements to take account of local factors such as pressure, temperature, the local tidal effects, magnetic and electrostatic fields of force, etc., etc., which may affect the length of our measuring rod. It may be that there are forces as yet unknown that are affecting our would-be 'rigid' rod. Surely it would be important for the purposes of distance measurement to know the disposition of these effects and forces and to know the position and orientation of our rod with respect to these fields of force and temperature, etc.'

Well, of course, that is so. But there is an enormous difference between calculating the length of our rod as a result of measurements made on conditions local to the rod, and calculating the length of our rod as a result of measurements made on the position of the rod with respect to some other point or points. We might, of course, come to know the conditions local to the rod by first making a plot of the conditions throughout some large area and *then* measure the position of the rod to determine the conditions local to the rod. That is, measuring the position of the rod may be a means to an end. But all we would *need* to know are the local conditions. And this would be so even if we were willing to countenance the existence of what Reichenbach has called 'universal forces'—forces which affect all substances equally *vis-à-vis* the length of the objects composed of the substances. The measurement of those factors which will affect the length of the rod is part of the measuring system. Part of the measuring system also is formed by calculations, via appropriate formulae, of the effects of these factors on the length of the rod.

Consider another example which does not involve the use of rigid rods, namely the use of a radar. A radar is not a measuring system. It is merely a measuring device. The measuring system is the way in which a type of radar is used. No mention need be made of any particular object to define a measurement system. The distance between any two points that the radar measuring system yields should be completely independent of the position and orientation of any radar used to implement the system. A measurement system is something of which there can be an infinite number of implementations *all* of which yield the *same* metric, that is, all of which yield the *same* distance between any two points at any time.

The restriction that we have been discussing may be put thus: the distance at any one time between any two spatially separated points *A* and *B* is independent of the position of the origin of, and the orientation of, any frame of reference or pointing system which is arbitrarily chosen for the purposes of spatial reference.

Let us call this principle the principle of independence of reference. Clearly, this principle is a considerable restriction on what counts as a distance-measuring system. But systems which meet this criterion may nevertheless fail to meet the other criteria mentioned on pages 51 and 52. For example, if the system yielded a distance of zero between two points which we knew to be physically distinct by virtue of some physical property that the one had and that the other lacked, then we would have to discard the system. It is for such a reason that we do not count such things as difference in temperature as a distance measurement. For two *different* points can have the *same* temperature.

Thus physical distance-measurement systems are certainly not things that can be chosen while sitting at an office desk, nor can they be convened upon by committees of scientists. That there are distance-measuring systems of particular kinds or, for that matter, that there are distance-measuring systems at all are theories about the nature of the universe. And such theories are to be tested in the laboratory and in the field.

But this does not answer the question of whether or not we can agree to use any one of an infinity of measuring systems given that we have found at least one. All we have shown is that an infinite number of congruence conventions, which can be generated from any one measurement system, are not in themselves measurements systems. But infinity is a funny number. One can subtract infinity from it and still be left with infinity. So there might still be an infinity of congruence conventions which would count as measurement systems for any one measurement system we could generate.

Let us return to the example of the rigid rod measuring system in use on the flat table. There we discounted the congruence convention whereby the length of the rod was counted as $1 - \dfrac{r\sin\theta}{2(1-r)}$ because it did not satisfy the principle of independence of reference. Could we describe the 'changing' length of the rod on the table by some means other than by making reference to the centre of the table? (Remember, the 'r' in our formula was the distance of the centre of the rod from the centre of the table.) Let us say that by some very rare chance there happens to be a magnetic field of force throughout the surface of the table such that the direction of the magnetic flux at any point on the table is always in a direction from the centre of the table to that point, and the magnitude of the magnetic flux density at any point is exactly equal to the distance of that point from the centre of the table. That is, if the magnetic flux density is B, then $B = r$. We could then specify our new measurement system, by saying that the length of the rod is $1 - \dfrac{B\sin\theta}{2(1-B)}$ where θ is now regarded as the angle which the rod makes with the direction of the magnetic flux. This new system now satisfies the principle of reference-frame independence. One no longer has to know anything about one's position or orientation on the table in order to apply the system. What one does need, however, is a device for measuring magnetic flux density and a magnetic compass for locating its direction. These, together with the rod, would constitute the apparatus needed to apply the system.

Let us now no longer assume that the magnetic flux density B is equal to r, nor that the direction of the flux is in a radius from the centre of the table. We could still use the formula as before. Of course, the distances it would yield would now be different. Also, we could vary this formula in such ways as to yield an infinity of different measuring systems. For example, we could say that the length of our rod was

$$1 - \frac{B^2 \sin\theta}{2(1-B^2)}$$

or $\quad 1 - \dfrac{B^3 \sin\theta}{2(1-B^3)}$

or $\quad 1 - \dfrac{B^4 \sin\theta}{2(1-B^4)}$

and so *ad infinitum*.

In short, given some quantity which has some particular magnitude for any particular point (this is what is usually known as a field), we can arbitrarily invent any number of distance-measurement systems which meet the restrictions so far laid down.

'But solid rods don't vary their length in a magnetic field according to *any* old formula', it may be objected. 'How a rod varied its length under such circumstances would depend for one thing, on the substance of which the rod is composed. Further, no substance would obey the laws of magnetic contraction that have so far been proposed. Magnetic fields do not cause changes in length in that way.'

To take the first point: it is true that the variation in the length of a rod in a magnetic field depends on the substance of which the rod is composed. It follows from this that the specifications of our measuring systems are incomplete. For any one of them would give us different answers for a particular distance, when different materials were used for the rods. Iron contracts in a magnetic field. Wood does not (or at least not noticeably). This leads us to another axiom of distance to add to our list: between any two points A and B there is one and only one distance. I shall call this the *axiom of functionality*. That is, a measuring system must determine a function from pairs of points to unique distances or, less technically, a measuring system has not been defined unless, for any two points at a particular time, the system as defined determines a unique distance. Given that one is operating the system as defined, it should not be physically possible for different implementations of the system to yield different results.

Very well, let us specify some particular substance for our rod, say pure platinum. *Now* there will not be any changes in the substance from rod to rod since, by virtue of our new measurement system, all measurement rods must be made of the same substance. It may still be objected that platinum obeys none of the laws of magnetic contraction proposed—let alone all of them. But if our distance-measurement system *defines* the length of the platinum rod as being $1 - \dfrac{B \sin \theta}{2(1 - B)}$, say, then that is what it is—as far as *that* distance-measurement system is concerned. It is not that the contraction of the platinum rod is to be explained as being caused or forced to occur by the presence of the magnetic field, any more than the presence of the magnetic field is to be explained as being caused by the contraction of the rod. It is just that this metric is determined by the rod and the field in concert.

'But this is intolerable', the physicist may cry. 'Here we have would-be quantities changing concomitantly without us being able to say that the first *causes* the second or vice versa. What is a physical quantity, if not a quantity a change in which requires causal explanation?'

The point is that changes in *lengths* of substantial objects do require explanation—an explanation in terms of forces which are changing the internal structure of the substance. If there were no such change in the internal structure, there would be good reason to believe that no change in length of the substantial object had occurred.

Thus, when a piece of iron changes its length as a result of being placed in a strong magnetic field, one can actually observe small domains of the crystals changing shape, shrinking, and growing. One can use *moiré* fringe techniques to observe the slipping of crystal imperfections that are associated with the stretching of metals under elastic stress. Observations of the so-called Brownian motion of tiny dust particles can be indicative of a change in mean free paths of the molecules of a gas under pressure—and so on.

The restriction on distance-measurement systems that we need here is this: any changes in the length of any substantial object are changes which accompany changes in the internal structure of the object which, in turn, are caused by forces local to the object and acting upon it. I shall call this principle the *causal principle of length measurement*.

One of the ideas involved in the causal principle which needs a little investigation is the idea of a structural change. What are we to count as a structural change? The notion is a familiar one, and paradigm cases of structural change can easily be cited—for example those cases mentioned earlier. But it is a borderline case which gives us difficulty. The case is that of some substantial object which increases (or decreases) its every dimension by some factor and whose every part also increases (or decreases) in every dimension by the same factor. Every part, and every part of every part, retains its orientation with the whole—there is simply an over-all and proportional increase (or decrease) in size.

Of course, few people, if any at all, believe in the existence of such an object or believe that such an object ever did or ever will exist. Most people believe that matter is composed of very small particles whose natures do not change at all while the particle remains in existence. It was thus that physicists thought of atoms in the nineteenth century and before, and it is thus that physicists now think of their sub-atomic particles. The ultimate building blocks of matter are deemed *not* to alter in any of their properties including their size. The size of larger objects may alter by virtue of the distance between these atomic or sub-atomic building blocks becoming greater—never by virtue of the size of the sub-atomic particles becoming greater, along with the size of everything else.

The platinum rod of our earlier example did behave that way, at least with respect to the particular metric of that example which was based on the platinum rod itself together with the local magnetic field. But this metric was not that of a measurement system according to the causal principle of length measurement. For although the change in would-be 'length' that occurred could be attributed to a change in a local factor, namely the magnetic field, it is false, on the assumption of that example, that this or any other local factor *acted on* the platinum thus *causing* the change in length.

Thus it would appear that we can allow our borderline case— namely an object whose every part increased in size by the same ratio—as a case of structural change, without our being unduly generous with respect to what counts as a distance-measurement system.

The causal principle of length measurement concerns changes in the length, or the lack of such changes in a single body between two different times. There does not seem to be any reason, though, why one should not extend the principle as criterion for equality of distances *through physical objects* generally.

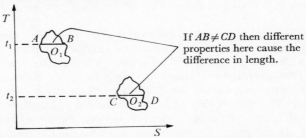

FIG. 13. *The extended causal principle of length measurement*

As shown in Fig. 13, let O_1 and O_2 be any two objects which may or may not exist at the same time or at the same place. Let t_1 and t_2 be two instants of time such that O_1 exists at t_1 and O_2 exists at t_2. Let A and B be points such that, at time t_1, both A and B lie on the surface of O_1, and all of the points between A and B lie within the object of O_1. Let C and D be points such that, at time t_2, both C and D lie on the surface of O_2, and all the points between C and D lie within O_2.

Then the distance between A and B at t_1 is equal to the distance between C and D at t_2 *unless* there exist differences between the properties of the object O_1 at time t_1 along the line AB and the properties of the object O_2 at time t_2 along the line CD, such that these differences in properties are causally related to the differences in the distances.

I shall call this the *extended causal principle of length measurement*.

The physicist's motivation for restricting length-measurement systems in accordance with this principle is the same as the motivation for restricting length-measurement systems in accordance with the original causal principle of length measurement. The origins of the motivation lie in the physicist's desire to find physical qualities and quantities with respect to which 'nature is uniform' or, in other words, with respect to which 'like causes produce like effects', as the position is often loosely described. Nelson Goodman in his book *Fact, Fiction and Forecast* (2nd edn., Bobbs-Merrill, New York, 1965), p. 79, differentiates between well-behaved predicates which are 'admissible in law-like hypotheses' and ill-behaved predicates that are not. The well-behaved predicates are those which have occurred frequently in hypotheses which have yielded many successful predictions in the past. By virtue of this, the well-behaved predicates thereby become 'well-entrenched' in the language. Obviously, the physicist who, *qua* physicist, will be interested in prediction and causal explanation will be seeking for qualities and quantities whose corresponding predicates will be well-behaved and which will thereby deserve to become entrenched in the language. Thus would-be measurement systems which yield 'lengths' that are known to be *ill*-behaved will be rejected for systems that are at least not yet known to yield such 'lengths'. The extended causal principle of length measurement is a minimum guarantee that the lengths yielded by a measurement system may be well-behaved, that is, that they may turn out to be likely candidates for causal hypotheses which will prove useful for prediction and causal explanation, for what the principle guarantees is that differences in length will be accompanied by differences in *other* physical properties.

If what I have said about the restrictions that physicists place on proposed measuring systems is correct, then it makes sense to believe that there is a distancing relation between pairs of points that is importantly unique. It may be that there are different measuring systems which yield this metric, but that is not to say that there need to be more than one such metric. What we believe this metric to be (assuming that there is but one) will depend on our beliefs about what causes what. But what this metric *is* will depend on what *does* cause what. However, it will not necessarily depend on the existence of any all-pervading aether or absolute space. In short, what I am claiming here is that it is consistent to believe in absolute distances and lengths without being an absolutist with respect to space.

3.7 THE NOCTURNAL EXPANSION PROBLEM

What I had just claimed at the end of the previous section has a bearing on an old problem in the philosophy of space. It is this:

Could all distances (including all lengths of substantial objects and their parts and all the distances between them) double in size overnight?

In the past, philosophers and physicists have tended to believe that if one answered 'yes' to this question one was thereby committed to an absolute view of space, and if one answered 'no' one was a relationalist. Why did people think thus? I think the idea was that there were but two possible views of spatial distance. The first was what we have dubbed the Newtonian view of spatial distance whereby if a body doubles in every dimension, it takes up eight times as much space—where the space is to be thought of in that Newtonian semi-substantive manner. A body that doubles in size soaks up space rather as a blotting-paper soaks up ink. Distances are absolute by virtue of these absolute quantities of space-stuff. The second view was that distances are relational and what this was supposed to mean in this context is that one cannot meaningfully refer to distances themselves, only distance ratios. A table has a length only by virtue of its bearing a certain distance ratio to a standard such as the standard metre-rod in Paris. Thus on this account one could say that all dimensions of all things, save the standard metre-rod in Paris, doubled in size; but one could not properly say that the metre-rod in Paris doubled in size as well.

However, there is at least one further alternative. A physicist may be a relationalist in the much weaker sense in which I have been using that term—namely one who believes that statements making reference to space can be reduced to statements which do not do so—and, at the same time, he may take an absolutist view of distances in the sense that one believes that physical distances are not merely a function of more or less arbitrarily adopted measuring systems.

So, now, to answer the question with which this section began: it *is* logically possible for everything to double in size overnight, for it is *logically* possible for all physical distances to be caused to change during the night by some factor that propagated itself throughout the universe, *causing* the increase in dimensions. A fantasy? Indeed; for we do not believe that such an event is *physically* possible. Vast accelerations of great masses would be necessary and we do not believe in the existence of forces necessary to cause such changes in motion. The sizes of all atoms and sub-atomic particles would have to change. Many laws of physics that we believe at least to approximate to the truth would in such a world be nothing like the truth.

'But how would one know, in a world in which it *were* true, that everything had doubled in size?' some philosopher or physicist may ask. To those with only a moderate training in philosophy or physics, that question may seem rather irrelevant in this context—and I

would agree that it is. However, there are those who would believe
that if that question cannot be answered, then the statement that
everything has doubled in size is meaningless. But it is very easy to
answer the question, for if I am right in my assumption that everything has not, in fact, doubled in size, then a world in which it *is*
true that everything has doubled in size is a fantasy world and, of
course, in science-fiction one may have one's cognitive beings relating
perceptually and hence cognitively to the world in any way one
chooses. Let them all *hear* the universe double in size.

To sum up, on a rather anti-climactic note, the nocturnal expansion problem, which has for decades been thought of as pivotal in
the philosophy of space by many philosophers, is far from central,
and what importance it has is relative only to a rather narrow and
naïve type of relationalism.

3.8 SOURCES AND HISTORICAL NOTES FOR CHAPTER 3

A beautifully written text covering the mathematical background
to this section is *Mathematics, its content, methods and meaning*, vol. 3,
edited by A. D. Aleksandrov, A. N. Kolmogarov, and M. A.
Lavrentev, and translated by K. A. Hirsch (M.I.T. Press, Cambridge, Mass., 1969), Chapters XVII and XVIII. Also, Bas van
Fraassen's book, *An Introduction to the Philosophy of Time and Space*,
mentioned earlier, is again very useful, especially his brief yet comprehensive survey of the history of the relationships between the
various branches of geometry, and his lucid introductions to those
concepts peculiar to geometrical thought.

Adolf Grünbaum's *Philosophical Problems of Space and Time*
(Routledge & Kegan Paul, London, 1964) is an invaluable source
book and is philosophically stimulating. This book contains,
perhaps, the most comprehensive treatment of the philosophical
problems of geometry in this century. However, Grünbaum's
approach is not without its critics, some of whom would defend
spatial absolutism against Grünbaum's relationalism; for example,
see Hilary Putnam's 'An Examination of Grünbaum's Philosophy
of Geometry' in *Philosophy of Science*, the Delaware Seminar, vol. 2,
edited by B. Baumarin (Interscience, New York, 1963), and
Grünbaum's 'Reply to Hilary Putnam's "An examination of
Grünbaum's Philosophy of Geometry" ', in *Boston Studies in the
Philosophy of Science*, vol. V, edited by R. S. Cohen and M. W.
Wartofsky (Synthese Library, D. Reidel, Dordrecht–Holland,
1968), and John Earman's 'Who's Afraid of Absolute Space?' in the
Australasian Journal of Philosophy, vol. 48, no. 3 (December 1970).

These articles include extensive bibliographies. All these authors seem to assume that the existence of absolute distances entails the existence of absolute space. It is this assumption that I have disputed in the last section of this chapter.

Poincaré's philosophy of geometry may be found in his *The Foundations of Science* mentioned earlier.

For further reading with respect to Bridgman's operationalism, see his *The Logic of Modern Physics* (Macmillan, New York, 1927), especially Chapter I.

For a brief statement of Karl Popper's philosophy of scientific method, see his 'Personal Report', in *British Philosophy in the mid-century* (Allen & Unwin, London, 1957), edited by Mace.

A work of classic importance in this area that must be mentioned is Hans Reichenbach's *The Philosophy of Space and Time* (Dover, New York, 1957).

Also of interest is Brian Ellis's very clear *Basic Concepts of Measurement* (Cambridge University Press, Cambridge, 1968), especially Chapter II, where he criticizes the sort of operationalism discussed in section 3.1 of this book, and claims that quantity concepts such as length are 'cluster concepts' in the sense of that term introduced by Douglas Gasking in 'Clusters', *Australasian Journal of Philosophy*, vol. 38, no. 1 (1960). As opposed to operationalism, there is the view which Ellis calls 'naïve realism' which is the view that a quantity 'exists, so to speak, before measurement begins. The process of measurement is then conceived to be that of assigning numbers to represent the magnitudes of these pre-existing quantities; . . .' My own position as outlined in this chapter has been slightly closer to naïve realism than the position which Ellis adopts.

CHAPTER 4

Time, Space, and Space–Time

4.1 SIMILARITIES AND DISSIMILARITIES BETWEEN TIME AND SPACE

Assuming that a relationalist programme is viable for space, how about time? The myth (if it is a myth), which relationalism attempts to eradicate from our ideas of space, is that at the back of, or somehow holding or supporting or providing room for, substances like water, air, soap, and so on, there is the 'substance' which is the daddy of them all—space.

A corresponding myth about time is that time is the daddy of all processes. It is the 'process' by virtue of which all other processes take place. There is a hymn which includes the words:

> Time, like an ever-flowing stream,
> Bears all its sons away.

The metaphysical convictions which lie behind these words are common to most of humanity. Yet these convictions are not without their paradoxes. If time flows, then it must be flowing at a certain rate. Now rates of flow are always with respect to time. So if time flows, it must be flowing with respect to itself. But nothing can flow with respect to itself. So time doesn't flow.

Arguments such as this have, throughout this century, convinced many people that there is no such thing as time in the sense that time is something that flows and is absolute, in the sense that its existence is not dependent on the existence of any other thing. Nevertheless they have felt that statements referring to time or using temporal concepts do usually convey useful and often important information. They have therefore chosen to embark on a programme of reduction usually in terms of relations between events. So the absolutist–relationalist dichotomy in the philosophy of space is reflected in an absolutist–relationalist dichotomy in the philosophy of time.

One important direction this relationalism has taken is in an attempt to liken time to space. Space is not like a flowing stream at all.

It is true that time is like space in many respects and in many more respects than is commonly realized. One of the reasons that some of the respects in which time is like space are not immediately obvious is that we often have words with which to describe some aspect of the temporal dimension which cannot be properly used to describe the *same* aspect of the spatial dimensions and vice versa. Thus 'instants of time' is correct English but 'instants of space' is not. Sometimes the language is biased in other respects also. For example the word 'speed' means the distance moved per unit time, but there is no single word for the reciprocal of this quantity, namely the time used per unit of distance moved.

One important point to be made here is this. If one finds a property that time can have which space cannot have, it does not follow that one has found a difference between time and space. Assuming that the 'can' and the 'cannot' are used to assert the absence or the presence of a contradiction respectively, all that one has found is a difference in the meanings of the words 'time' and 'space'. For example, the following argument is invalid.

Nurses can be male.

Females cannot be male.

Therefore nurses are different from females.

If the conclusion means that some nurses are different from females, it may be true, but it would not follow from the premisses, for it is consistent with the premisses that all nurses are female. (It's just that no males have happened to venture into the profession, say.) If the conclusion means that all nurses are different from females, then the conclusion is obviously false and since the premisses are true, the argument is invalid. Analogously, the following argument which is pertinent to recent philosophical literature on the mind–body problem is invalid.

Brain processes can be things of which no conscious being is aware.

Sensations cannot be things of which no conscious being is aware.

Therefore sensations are different from brain processes.

Because it is of the same type, the following argument would be invalid also.

Time can be a great healer.

Space cannot be a great healer.

Therefore time is different from space.

There are many ways in which modal words (words such as 'can', 'cannot', 'may', 'might', 'must', 'necessary', 'possible', and 'impossible') are used fallaciously to arrive at conclusions which assert the existence of distinctions which are not to be found in nature. For example, consider the following argument, once again taken from the mind–body controversy.

Necessarily all sensations are non-physical.

Necessarily all brain-processes are physical.

Therefore necessarily no sensations are brain-processes.

So far so good. One may not believe the premisses. But assuming that they make sense and are true, the conclusion follows. What does not follow from this conclusion, however, is the further conclusion that there are at least two different sorts of things to be found in the world—sensations and brain-processes.

From the fact that we are *able* to make distinctions within our language, it does not follow that mother nature follows suit. Depending on what other beliefs one has, and on what other evidence is available, it might be more reasonable to conclude that, strictly speaking, there are no sensations, but there are things that have all the properties essential to being a sensation with the exception of being non-physical; and further, that these things are brain-processes. Of course, if this sort of thinking were widespread, the meaning that people gave to the word 'sensation' would probably be modified so that the word delineated more useful boundaries. What would then be called sensations would no longer be deemed to be non-physical.

So it is with the argument:

Necessarily time flows.

Necessarily space doesn't flow.

Therefore necessarily time is not space.

For, as we have already surmised, it may be that the belief that there is something that flows in the sense that time flows is a mythical belief. It could be that we are a kind of creature whose perceptions of the world, as it happens to be in our vicinity, are such as to give us the illusion of some sort of procession of time or movement through time. Let us assume that, somehow or other, we came to know that this was the case. Even if it were necessary in our language that anything that was time had to be something that flowed, we would have to believe, in all consistency, that there was no such thing as time or that time is unreal as some philosophers have said. We may still believe that there is some sort of fourth dimension over and above the three spatial dimensions, but strictly speaking it would not be a time dimension. At this stage we would probably begin to twist our language a little, start calling this dimension time as before, and start denying that it is necessary that time flows. Indeed we would probably start asserting that it is contingently false that time flows. We shall return to this matter of time flow in Chapter 5. Meanwhile let us look at other ways in which it is possible to go astray *vis-à-vis* arguments purporting to show that time is different from space.

One such way is to put forward an argument to the conclusion merely that time is *numerically* different from space, while believing that you have found a *quality* in which time differs from space. For example, suppose *B* believes that time is just another dimension over and above the three spatial dimensions, but not qualitatively different from them. C may try to 'disillusion' him by pointing out that time is an aggregate of instants, whereas space is not.

B might then ask C what he thinks instants are, and C replies, let us suppose, that they are temporal infinitesimals. But B, who is not yet convinced that time is anything but a fourth dimension, will picture temporal infinitesimals as three-dimensional surfaces satisfying an equation of the form

$$t = k$$

where t is the time at any point in the surface and k is a constant. What he will still be wondering is whether or not this will be any different from a three-dimensional 'surface' generated by the equation

$$x = k$$

where x is the distance at any point in the 'surface' from any arbitrarily chosen spatial plane. If it is not, then there isn't a difference between temporal infinitesimals and other spatio-temporal infinitesimals.

All that C has shown, as far as B is concerned, is that one of the dimensions of space–time is numerically different from the other three. C has not shown that it is qualitatively different in some particular way.

Let us describe C's performance as a case of *special pleading* and let us call concepts, by virtue of which this kind of special pleading is possible, *special pleaders*. Thus the concept of an instant is a special pleader for discriminating time from spatial dimensions.

Here is a simple case of special pleading:

David is different from Tom in so far as David is 'davidfooted' whereas Tom is not. What does 'davidfooted' mean? It means having at least one of David's feet.

The concept of being davidfooted is a special pleader. In trying to sort out where temporal aspects of the world, including time itself, differ from the corresponding spatial aspects of the world, the biggest problem is to spot the special pleaders.

The example about time having instants did bring up one important difference between time and space, namely that there is but one dimension of time whereas there are three dimensions of space. Taking this into account then, as we did in that example, the question to ask is not

'What is the difference between time and space?', but rather,

'What is the difference between time and a spatial dimension?'

Given all the linguistic pitfalls of which we have been talking in this section, this is not an easy question to answer, if we think with words. Indeed, this is one of the few areas where it helps to use a

little picture thinking. To illustrate this, consider the following proposition:

A car is in a garage at time t_1, is driven out and subsequently is
not in the garage at time t_2, and is driven back into the garage at
time t_3.

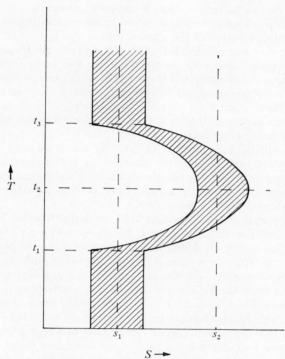

Fig. 14. *A car which has been driven away from some
position and then driven back into that position*

Can there be an analogue to this story in which the time dimension
is swapped for one of the spatial dimensions? Thinking in words,
one may come up with: 'Oh, no. The car can be at the same place,
namely the garage, at two different times, but nothing can at one
time be in two different places' or:

'This is a story of an object moving back and forth in space. You
cannot have an object moving back and forth in time.'

But let us plot the situation given in the example on a graph.
The position, S, in some chosen spatial dimension is plotted along
the horizontal axis, and the time, T, is plotted vertically (see
Fig. 14).

In Fig. 14, the shaded area is the car. It is at s_1 at times t_1 and t_3 but not at the intervening time t_2. At time t_2 it is at s_2. The width of the shaded area indicates the length of the car.

The question is: Can we rotate the shaded area, and still make sense of the picture? If we can, this particular example has not given us reason to believe that a temporal dimension is different from a spatial dimension. Figure 15 is the corresponding space–time diagram after rotation of the shaded area.

Can we give a sensible description of the events portrayed by this diagram? Quite easily. At time t_1 there are two large objects, one which covers the point s_1 and the other of which covers the point s_3. There is an intervening point s_2 which neither of them covers at time t_1. Shortly after t_1, the objects suddenly shrink and move towards one another. At time t_2 they have fused into one object and shortly afterwards this object vanishes.

This could be a description of drops of water on a hot griddle. The point to note is that the description corresponding to this diagram does not seem to conflict at all with the statement that an object cannot be in two places at once or that an object cannot move back and forth in space.

Actually it is not strictly true that an object cannot be at the same time at two different places. My hands are both parts of the one object, namely my body, and they are in different places. It is not even true that there cannot be a gap between different parts of a body in some given dimension of space. I can assert quite truly that in the East–West direction there is a gap between my hands at this moment which includes no other part of my body. The surfaces of objects can be concave as well as convex. What is meant by the assertion that an object cannot be in two different places at the same time is that it is possible, *at any time*, to trace a curve from any *point* within an object to any other *point* within that object without leaving the object. And, of course, in this sense it's true that an object cannot be in two different places at the same time. But this is true by virtue of the way we happen to count objects. The italicized expressions, '*at any time*' and '*point*', show that this principle of individuation is a special pleader for a difference between space and time. If we were better at prediction and retrodiction than we are, we would probably find the following principle of individuation of physical objects more convenient than the one we *do* use.

Here's the principle:

By 'point–instant' let us mean a point in space at some particular time. Then, given any two point–instants within an object, there is a spatio-temporal curve from one point–instant to the other, such that every point–instant on the curve is within the object.

Given *this* principle of individuation, in which there is no special pleading, the events portrayed by the diagram of Fig. 15 could have been described in the same way as the events portrayed in Fig. 14 were described, with points replacing times and vice versa. That is, we could have said that Fig. 15 was a diagram of *one* object which at t_1 is at points s_1 and s_3, but not at the intervening points s_2, and that the object is at s_2 at time t_2.

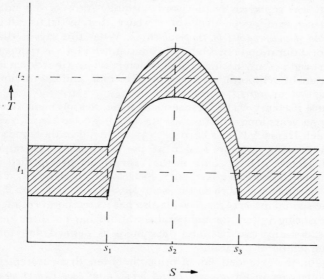

FIG. 15. *Two bodies run together, fuse, and then vanish*

In investigating this example, then, nothing has been found which differentiates time from a spatial dimension. However, something has been sorted out about how we count physical objects.

Richard Taylor has said that comparing space and time is 'intellectually stimulating and edifying to the understanding'. I think that what is so stimulating and edifying about it is that one so often learns something unexpectedly as in the example above. Also, one is led to envisage possibilities that one never envisaged before, and that is always exciting.

4.2 MEASUREMENT OF TIME, AND THE FROZEN UNIVERSE PROBLEM

In section 3.6 it was argued that the existence of absolute distances does not entail the existence of an absolute space. Similarly, and for similar reasons, I would claim that the existence of absolute temporal durations does not entail the existence of an absolute time. Any time-measuring system worthy of the name would have to satisfy a set of principles, called, let us say, the axioms of temporal duration, paralleling the axioms of distance. Indeed, all the axioms of distance enumerated on pages 51–2 could be taken directly as axioms of temporal duration also, given that the word 'distance' was replaced by the words 'temporal duration', and the word 'point' was replaced by the word 'instant'. Also corresponding to the principle of independence of reference, there is (let us call it) the *principle of independence of temporal origin*. What this says is that the temporal duration between any two instants t_1 and t_2 is independent of the origin of any dating system which is arbitrarily chosen for the purposes of temporal reference and is independent of how the dating system *orders t_1 and t_2*.

Thus, if Henry VII's life counts in some measurement system as being 52 years long when we use the existing calendar with respect to which Henry VII was born in 1457, it should in the same system count as being 52 years long if we use a calendar with respect to which Henry VII was born in the Year minus 26 and died in the Year plus 26 or a calendar with respect to which Henry died in the Year zero and was born in the Year plus 52.

The spatial *axiom of functionality* also has its temporal counterpart. A measuring system for temporal durations has not been defined, unless, for any two instants, the system as defined determines a unique temporal duration.

Again, the extended causal principle of length measurement has a counterpart in an *extended causal principle of time measurement* (see Fig. 16).

Let O_1 and O_2 be any two objects as before. Let s_1 and s_2 be two points such that s_1 lies within O_1 and s_2 lies within O_2. Then:

The temporal duration of O_1 at s_1 is equal to the temporal duration of O_2 at s_2, *unless* there exist differences between the properties of the object at place s_1 throughout the duration of O_1 at s_1, and the properties of O_2 at s_2 throughout the duration of O_2 at s_2, such that these differences in properties are causally related to the differences in duration.

As was the case with distances and space, it makes sense to believe that these restrictions on the measurement systems for temporal durations are satisfiable only by systems that yield the same metric for time. Again, what we believe this metric to be (assuming for the moment that there is but one such metric) will depend on our beliefs about what causes what, and again, what the metric *is* will depend on what *does* cause what. Needless to say, the existence of such an absolute metric would not imply that time is absolute in the sense that it is some sort of river-like process which is the aggregate or the source of all other processes.

One reason, however, why people regard time as the aggregate of all processes or change is the way we measure time. We use processes and changes in our measurements of time: the change in the positions of the hands of a clock, the change in the position of the stars relative to a meridian, the change in the length of a tallow candle as it burns, the process of sand running through a hole as

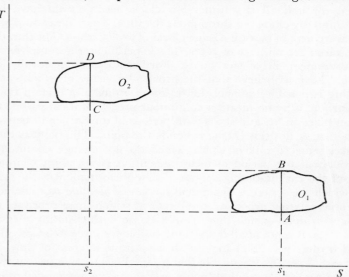

FIG. 16. *Illustrating the extended causal principle of time measurement*

in an egg-timer, and so on. If none of these processes were available, then we would have to use some other process. But what if there were no processes at all? What if the universe were completely frozen and nothing at all were changing? Then, of course, we would be frozen and unchanging also, and hence we would not be thinking or measuring either—nothing would be.

'Wait,' someone may object. 'All methods of measuring time utilize change. Whatever the method, if there is no change, the measurement of the amount of time elapsed is zero. And if all possible measurements of the elapsed time are zero, then surely the elapsed time must *be* zero. In which case there simply cannot be a frozen universe for some non-zero time; and, of course, to say that something is frozen for zero time is to say that it is not frozen at all.'

Now it is probably true that all methods of measuring elapsed time utilize change. But it is the second premiss that is confused.

What the objector believes is that the clock that is being used to measure time has to be running during the time that is being measured, and time measurement always has to take place during the time that is being measured. But neither of these propositions *has* to be so. The second proposition is not true even in this universe— let alone imaginary ones. What is true is that if an observer is going to *know* that some time has elapsed, and how much, he has to make sure that some change has taken place. But he needn't have lived through all that time. His clock needn't even have been continuously changing throughout the time. Take, for example, the measurement of past time using clocks of fossil carbon. The durations measured are millions of years. The experimenter accomplishes his measurement in a few hours. Furthermore, modern physicists would have us believe that the process by virtue of which carbon dating is possible, namely the fission of atoms of a certain isotope of carbon, is by no means a continuous process, but rather one that takes place in fits and starts with periods of quiescence in between. Whether or not this is true is beside the point. The point is that it makes sense to talk of clocks which do not change continuously throughout the period of measurement. Another point of interest about the carbon-dating system of time measurement is the extent to which the system is dependent on *theories about the past*, including theories relating processes to the *rates* at which those processes take place—without which there would be no carbon-dating measurements. So it is not at all *logically* impossible for there to be a world in which there was no change at all for a finite amount of time. It is not even logically impossible that an intelligent being in that world should come to know that such a period of time had existed or even was about to exist.

Some might object that, whatever system of time measurement was used by such an observer, it could not be one that satisfied the extended causal principle of time measurement; for the concept of cause which arises in this principle is one which is related to a causally deterministic universe: that is, a universe in which what goes on in the future relative to some instant is necessarily determined by what the universe is like at that instant, and which also is such that its state at some instant is causally sufficient to determine the history of the world after that instant.

Now let us assume that, during the time the universe is frozen, it is in some state Φ. Then there will be instants when the universe is in state Φ, and which will be such that the universe will be in state Φ for some time, say Δt, afterwards. But consider the instant at which the universe is in state Φ but will remain in this state for a duration of only $\frac{1}{2}\Delta t$. This instant will *not* be succeeded by a period during which the universe remains in state Φ for a period of Δt. Thus the universe's being in state Φ at some instant will not be a *sufficient* condition for the history of the universe following that instant to be whatever it turns out to be. Like causes would not produce like effects.

But one may object that this conception of causal determinism is unnecessarily narrow. One can conceive of other mechanisms whereby the history of a universe may be determined.

Consider, for example, the following model for a universe. There are only two possible states that the universe can be in at any time. One of these states will be called the '1' state, the other will be called the '0' state. Let Φ_t be the state of the universe at time t and let $\Phi_{t-1}, \Phi_{t-2}, \ldots, \Phi_{t-n}, \ldots$, be the states of the universe at times $t-1, t-2, \ldots, t-n, \ldots$ There is only one law of nature applicable in the universe and it is this: $\Phi_t = 0$ if $\Phi_{t-3} = \Phi_{t-4}$, and $\Phi_t = 1$ if $\Phi_{t-3} \neq \Phi_{t-4}$. A diagram showing the history of such a universe is given in Fig. 17. In this particular universe, history repeats itself every fifteen units of time.

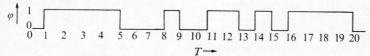

FIG. 17. *A two-state universe with mnemic causation*

The state of the universe at any instant is not sufficient to determine the future of the universe from that instant, but the future of the universe is determined just the same. The history of the universe for four units of time prior to any instant is sufficient to determine the history of the universe after that instant.

The example given is just one type of causation that is different from the type normally envisaged. But there are, of course, an infinity of variations on this theme. Interestingly enough, the universe in the example given is causally mnemic, or, to put it another way, in that universe, there would be action over a temporal interval. One of the differences in the history of the philosophy of space and time has been that people have long argued whether or not there could be action at a distance, that is over a spatial interval, but no one, with the sole exception, as far as I know, of Bertrand Russell, has ever considered whether or not there can be action over a temporal interval.* But the difference would appear to be an accident of academic history—not a result of any differences between the way we think of space and the way we think of time.

---

* Since writing this, it has been brought to my attention that J. J. C. Smart mentions this possibility on p. 296 of his *Between Science and Philosophy* (Random House, New York, 1968).

The point of discussing these imaginary universes is to investigate alleged conceptual differences and similarities between space and time. But such exercises often have a mind-broadening effect as well. Hypotheses that one used to regard as logically absurd become possibilities and, this being the case, avenues that have previously been closed off for research can be opened.

4.3 QUESTIONS OF TEMPORAL TOPOLOGY

In the last section a simple universe was described whose history, it was said, repeated at fifteen-unit intervals. In the diagram illustrating this universe (Fig. 17), the scale went from zero to twenty and there was every indication that this was only a little piece of the picture. The actual universe would go on *ad infinitum* in each direction. But why not say that instant number 17 is the *same* as instant number 2 and that instant 18 is the *same* as instant number 3 and so on? That is, why think of this example as a temporally infinite universe repeating itself over and over again, rather than a universe wherein time is finite but circular; where any instant of time will be in its own future and its own past? The total temporal duration of the universe would be fifteen units.

In section 3.2, questions were raised concerning the topology of space. The question now to be asked is: 'Does it make sense to say that time, like space, has topological properties?' On the assumption that the previous paragraph made sense, it would appear that it is sensible to say that time has topological properties, for in that paragraph we were raising the question of what sort of topology was exhibited by the mnemically causal universe of section 4.2. Is it a finite circle of 15 units of time, or is it an open-ended line of an infinite number of units of time?

The answer, I would claim, is that that part of the story has not been told. It is, as far as the description of that universe that has so far been given is concerned, an open question. In his *Philosophical Problems of Space and Time*, Grünbaum calls a universe that goes on and on *ad infinitum* into the past and into the future a temporally *open* universe. The other sort where, for each instant, all the instants of the universe are both in that instant's future and in its past he calls temporally *closed*.

Now not all people are as open-minded about temporal topology as I have been. Grünbaum has claimed, doubtless correctly, that many would wish to assert that any description of a temporally closed universe is inconsistent or 'that it is of the essence of time to be open'. Grünbaum himself does not make such a claim, yet, in discussing a universe similar to the one discussed above, he says that it is inadmissible to describe the time of that universe as topologically open and infinite in both directions. This interpretation is illegitimate in this case, he claims, '*since a difference in identity is assumed among events for which their attributes and relations provide no basis whatever*' (my italics).

The universe that Grünbaum is discussing is that of a single particle moving in a circular path without friction. The infinitely repeating temporally open model which Grünbaum disallows is shown in Fig. 18(a); that of the corresponding temporally closed model is shown in Fig. 18(b). (Of course, only one spatial dimension is shown.)

In the open-ended universe, the particle passes through s_1 at different times t_1, t_2, t_3, t_4, etc. In the closed universe, the particle passes through s_1 only once.

In disallowing the open-ended model Grünbaum takes himself to be resting his conclusion on Leibniz's thesis 'that if two states of

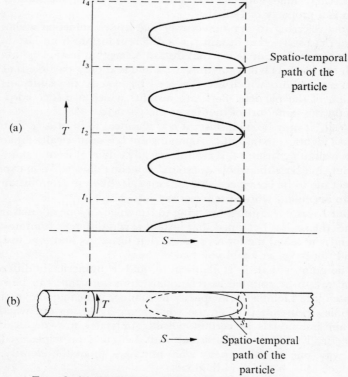

FIG. 18. (a) *Oscillating particle in a temporally open universe*
(b) *Oscillating particle in a temporally closed universe*

the world have precisely the same attributes, then we are not con-
fronted by distinct states at different times, but merely by two
different names for the same state at one time'.

The use or rather misuse of principles such as Leibniz's principle,
to restrict unnecessarily the conceptual freedom of theoreticians,
has been a feature of twentieth-century philosophy in particular,
but it has been common throughout the history of philosophy.
For that reason I would like to pause a while to examine in depth
this little philosophic misdemeanour of Grünbaum's, for though it
is fairly innocuous in its context, it is a good example of a fairly
widespread disease.

Leibniz's principle is often formulated thus:

Given any two numerically distinct things, then there is some
property held by one that is not held by the other.

The word 'property' here is used to include relations that an
object may hold with another object. Thus being the mother of
Kim is a property of Delilah.

Now referring to our universe of Fig. 18(a), someone wishes to
claim (let us say) that instant t_2 is a different instant from instant t_1.
What properties can he say that instant t_2 has that instant t_1 hasn't?
Both are instants when the universe's one and only particle is at s_1.
Both are instants which are preceded by exactly the same sort of
motion of the particle. Both are instants which are succeeded by
exactly the same sort of motion of the particle.

'Well,' it may be said, 'instant t_2 is identical to instant t_2. Instant
t_1 is *not* identical to instant t_2. Again instant t_2 is numerically different
from instant t_1. Instant t_1 is *not* numerically different from instant t_1.'

'But that begs the whole question,' someone objects. 'You cannot
expect me to be convinced that instant t_2 is different from instant t_1
by an argument whose premisses say just that.'

'But I was not arguing or trying to convince anyone of anything,'
comes the reply. 'I claimed that instant t_2 was different from instant
t_1, and you asked me for a property that instant t_2 had that instant
t_1 did not have, so I told you two of them.'

The point is that, if 'is identical to' and 'is numerically different
from' are to be counted as relational expressions that can be substituted into Liebniz's principle in the manner indicated above, then
Leibniz's principle is of no use whatsoever in sorting out whether
certain individuals in certain possible universes are identical or
numerically different. Consistent with Leibniz's principle used in
this way, one can always say what one likes about the identity, or
the lack of it, between individuals.

Very well. Let us not allow that being identical to something
and being numerically different from something can be properties

that a thing can have—at least for the purpose of this Leibnizian exercise. Can our champion of open-ended time now defend the difference between instant t_2 and instant t_1? Yes. For he could say that instant t_2 had the property of being *later than* instant t_1 whereas instant t_1 did not have the property of being later than instant t_1. Once again, anyone who believes that instant t_2 is the same instant as instant t_1 is not likely to have his mind changed by an argument whose premiss is that instant t_2 is later than instant t_1. He just won't believe the premiss either. But that is beside the point. The point is that the open-ended interpretation of our universe remains just as valid under our new improved Leibniz's principle as before. Further, the charge of question-begging in matters of identity can be levelled at any descriptive expression whatever. Nobody who is firmly convinced that B is the same thing as C is going to believe that B has a property which C lacks.

So let us cast aside the Leibnizian red herring and re-examine the structure of Grünbaum's argument. What it looks like is this. We are asked to assume that there is a universe V, with certain properties F, G, and H, say. (It doesn't matter for the moment what these properties are.) V's having properties F, G, and H 'provides no basis whatever' for the assumption that V has property J (where J is some further property). It is, therefore, 'illegitimate' to further describe V as having property J. Therefore, on the assumption that V is F, G, and H, V cannot have property J. Put more bluntly, the argument is of the form.

$$p \text{ does not entail } q.$$
$$\therefore \quad p \text{ is inconsistent with } q.$$

Compare:

'Tom is a man' does not entail 'Tom is married'
$$\therefore \quad \text{'Tom is a man' is inconsistent with 'Tom is married'.}$$

The argument is so blatantly invalid that one might suspect that more is going on than has been described. For example, someone might interpret Grünbaum's argument as a verificationist argument. What does that mean? The verificationist has turned up in previous sections. He is the man who insists on believing that if there is no way of finding out whether or not a statement is true then that statement is meaningless. A man reading Grünbaum's argument through spectacles of this sort might render it thus:

Given that V has properties F, G, and H, there is no way of telling whether or not V has property J. Therefore, given that V has properties F, G, and H, it is inadmissible to assert that V has property J (since 'V has property J' will be meaningless). But this will never do, even if we were to accept the verificationist's principle.

For if there is no way of telling whether or not V has property $\mathcal{J}$ (i.e. being temporally open), there is no way of telling whether or not V has the property of not being $\mathcal{J}$ (i.e. of being temporally closed) either. So it would also be inadmissible to assert that V is temporally closed. But this is what Grünbaum wanted to do.

Someone may say at this stage that what *should* have been said is that the descriptions of universe V as being temporally open *or* temporally closed are both meaningless. But even this would seem to be strange, because one would be using the verificationist's principle in such a way that whether what one said was meaningful or not would always depend on the story with which one prefaced what one was saying.

What is more, we know very well that there is no such universe of the kinds described either in this section or in section 4.2. The actual universe is much more complicated, for one thing. So we know that the descriptions of that universe are false, and will remain false no matter how those descriptions are extended— whether we say that it is temporally open and repetitive or temporally closed and non-repetitive. And presumably, if some statement is false, then the expression in which the statement is couched is meaningful.

On the other hand, were the verification principle to be couched in terms of what one could get to know in the universe being described, were it to exist, one would again be in difficulty. For if there were no intelligent beings in the universe, nothing could be known in the universe and hence any description of the universe would be meaningless; and if the description of the universe were fleshed in to include beings who could know things, then why not flesh it in to the extent where they could know whether or not instant t_2 was or was not the same instant as instant t_1?

As an alternative to these verificationist approaches, one might imagine that what someone may have in mind when using an argument such as Grünbaum's is some sort of analytic or explicatory reduction of statements involving temporal relations in accordance with which it would be necessary that, if a universe V had the properties it was alleged to have, then it would also be temporally closed. But as an analysis of normal linguistic usage, such a reduction would in this case seem rather bizarre, and if the reduction was regarded as an explication, one might ask why it was desirable to impoverish our language in this way. To be able to describe the universe either way with respect to temporal closure does not seem to lead to paradoxes.

Again, someone might imagine that the argument is based on some ontological or theoretical reduction of temporal relations which act as a hidden premiss in the argument. But this will never do, for ontological and theoretical reductions are applicable contingently to the universe that *does exist*. The existence of certain sorts of things and the contingent identity of properties in some imaginary universe has no bearing whatsoever on what *does* exist and what properties *are* in fact contingently identical, nor do the contingent facts of this world restrict in any way what contingencies are true in some imaginary world.

The motivation for bothering with this argument does not arise from an avid interest in strange imaginary universes. What I have wanted to make clear is that neither the use of Leibniz's principle nor the use of verificationist methodologies yield quick answers to problems of identity, such as the one we have been discussing, but including also such problems as:

Can there be a universe one half of which is a mirror-image of the other?

Can a universe be spatially repetitive?

Can time begin to flow backwards?

Neither Leibniz's principle nor verificationism can turn possible worlds into impossibilities. Nor can they act as a guide for opting for one of the possibilities that do obtain.

One more word about these cases for those familiar with the literature. It is often thought that a relationalist would have to answer 'no' to the above questions. Take the case of the mirror-image universe, for example. It is argued that, if a mirror-image universe is logically possible, then an effect and its mirror-image counterpart have all their monadic and relational non-spatio-temporal properties in common. If someone envisages this possibility, he has rejected a relational view of space and time, for he holds that spatio-temporal relations cannot be reduced to non-spatio-temporal relations.

Now the relationalist we have been discussing throughout this book is not one who wishes to provide reductions for spatio-temporal relations, but rather for space and time. In fact, I have never met or heard or read of anyone, apart from those with verificationist tendencies, who has felt any *need* to provide reductions for all spatio-temporal relations, though I do not deny that there may be some, for example, E. J. Zimmerman, mentioned in section 1.4, who think that a theoretical reduction of all such relations may be possible. The only spatial and temporal relationships that one must be able to retain to enable one to envisage the possibility of the kinds of universe mentioned above are the relationships of being 'spatially distant from', 'temporally distant from', 'spatially coincident with' and 'temporally coincident with'. Do these relationships offer some difficulty for the kind of relationalist who simply wants to be rid of 'space' or 'time'? Perhaps someone may wish to argue thus:

'To say that *B* is spatially distant from *C* is analytically equivalent to saying that *B* and *C* are in different places or positions or at different points. So someone who claims that *B* is spatially distant from *C* is committed to an ontology which includes places or positions or points. And anyone who is committed ontologically to spatial places, positions, or points is committed to space, for space is simply the aggregate of these entities.'

We have already dealt with the difficulties for our relationalist with respect to spatial points in section 3.5. It is the argument from the first sentence to the second that I wish to take issue with here. For if it is true that '*B* is spatially distant from *C*' is analytically equivalent to '*B* is in a different place or position from, or is at a different point from *C*', then this very analytic equivalence would provide the basis for an analytic reduction of the terms 'places', 'positions', and 'points' in contexts of this sort, using expressions which make no reference to places, positions, or points.

The motivation for the sort of relationalist we have been considering throughout this book, is a motivation to resolve certain paradoxes about space and time. But I know of no paradoxes that can be generated by assertions to the effect that there are objects which are not spatially or temporally coincident.

4.4 MORE DIFFERENCES BETWEEN TIME AND SPACE

In this chapter so far, it has been shown how like time is to space in many ways, to the point where many philosophical problems about time have their counterpart in problems about space. Yet a glance down the list of properties of space would convince many people (if they needed convincing) that time is very different from space. We don't believe that time has a permittivity or magnetic permeability, or that time is transparent to light, or that time is a good or poor conductor of electricity. Many of these would-be propositions don't even seem to make sense. But that should make us wary. For neither does it make sense, on the basis on which 'david-footed' was defined in section 4.1, to state that somebody who isn't David is davidfooted. These properties of space may be special pleaders. If time is not just another dimension, then there must be laws of nature involving time which do not have a counterpart in which the reference to time cannot, without generating a falsity, be replaced by a corresponding reference to a spatial dimension, and, in those cases where reference to space occurs in the law, one of the spatial dimensions is replaced by time.

One such alleged law which will receive closer attention in Chapter 6 is the Second Law of Thermodynamics. This states that there is a property of any causally isolated system called its entropy which always increases with increasing time. Never mind for the moment what entropy is. Nobody ever alleges that a causally closed system always increases its entropy as the distance increases in some given direction. Another such theory is Maxwell's theory for electromagnetic radiation, a consequence of which, as was mentioned earlier, is that electromagnetic radiation has a constant speed c_0 in empty space. This is consistent of course, with its speed being zero *in a direction perpendicular to that in which the wavefront happens to be travelling*. What is the case is that the resultant of the velocities in each of three mutually perpendicular directions must yield a speed of c_0. Depending on our choice of mutually perpendicular axes, the velocity of a particular ray of electromagnetic radiation could be anything from zero to c_0 in the direction of one of the axes. That is, the rate of change of distance with time of some part of the wavefront of any particular ray of electromagnetic radiation in some chosen direction will be either zero or c_0, or will be between zero and c_0. But this in turn implies that the rate of change of time used with respect to the distance travelled in our chosen direction is either infinite, $1/c_0$ or some value lying between these two; in any case, some value greater than zero. And herein lies a difference between time and any dimension of space. The rate of change of distance in some direction with respect to time of a part of a wavefront of electromagnetic radiation may be zero; but the rate of change of time used with respect to the distance travelled by such a part of a wavefront in some direction can never be zero.

But it seems objectionable that one has to go to such esoteric sources to find differences between time and spatial directions. The important differences seem to be immediately obvious to all: the future seems to approach and the past to recede regardless of our spatial movements or whether we are moving at all; yet places in a certain direction approach us only if we move towards them; and our moving takes time. Spatial movement seems to be within our control, but our movement through time seems to be inexorable. We approach the end of our existence whether we like it or not. There is no reversal. There seems to be *a preferred direction of time from earlier to later* which is not arbitrary in the way that 'to the left' and 'to the right' are arbitrary. We change what was to the left to something that is to the right simply by turning through an angle of 180°. But we do not seem to be able to involve ourselves in some sort of spatio-temporal revolution to turn what is past into something that is in the future. People may have feelings of nostalgia, sorrow, or glowing happiness when they remember past events, but they do not plan for the past. It is silly, it is often said, to worry about the past, to cry over spilt milk as the saying goes. But no one thinks it silly to concern oneself over whether or not the milk *will* spill. Something may be done to *prevent* it from spilling.

John Earman has argued that many unnecessary philosophical problems about space and time arise out of the fact that philosophers tend to treat space and time as two separate entities, instead of the one entity space–time.

'Space', he says 'is not a given entity like the earth; it is space–time that is given, and space must be sliced off from space–time. And if there is one way of doing the slicing, there are many.'

I would guess that most modern physicists would agree with Earman on this point. Nevertheless, most physicists would also agree that the ways we *do* normally slice off space from space–time or, more strongly, the ways in which we *can* slice off space from space–time, are such that the spatial dimensions so sliced off will have very different properties from the temporal dimension. Indeed, in Einstein's Special Theory of Relativity, which could be regarded as the first thoroughly articulated theory of space–time in Earman's sense, it is important to distinguish between the temporal dimension of some reference-frame in space–time, and the three spatial dimensions. Further, though what counts as the temporal dimension can vary from one reference-frame to another (from one observer to another, if you prefer), the extent of this variation has its limits, according to the theory.

As observers in space–time, we find that differences between the temporal dimension and the spatial dimensions seem immediately obvious. This has already been said. It has been said also that these differences are very important to most of us. It is interesting, therefore, to attempt to analyse the way we think about these differences, to try to state as explicitly as possible what differences we actually experience, and to try to explain these experiences with available physical theories.

The outstanding differences between the way we ordinarily think of time as opposed to some spatial dimension are:

(*a*) our attitude towards what is present or *now* as opposed to what is *here*, particularly with respect to our feelings that what is now has a preferred status of existence to what was or what will be; and

(*b*) our belief that this present somehow moves in a preferred direction in time from what we call earlier to what we call later.

Chapter 5 will be concerned primarily with the first of these considerations. Chapter 6 will be an investigation into what reasons we have, apart from our intuitions, to believe that the temporal dimension exhibits any asymmetry.

4.5 SOURCES AND HISTORICAL NOTES FOR CHAPTER 4

The argument with which section 4.1 begins appears in C. D. Broad's *An Examination of McTaggart's Philosophy*, vol. 2, Part I (Cambridge University Press, Cambridge, 1938).

The material in the latter pages of section 4.1 was largely stimulated by the first half of Chapter 6 of Richard Taylor's book *Metaphysics* (Prentice-Hall, Englewood-Cliffs, N.J., 1963). The chapter is entitled 'Time and Becoming'. The views expressed by Taylor in this chapter are also expressed in his 'Spatial and Temporal Analogies and the Concept of Identity', printed in *The Journal of Philosophy*, vol. 52 (1955). The paper, so Taylor claims, was inspired by the concluding chapter of Nelson Goodman's *The Structure of Appearance* (Harvard University Press, Cambridge, Mass., 1951).

The frozen universe problem discussed in section 4.2 has a very long history indeed. Aristotle in his *Physics* defines time as a 'number of movement'. Thus no movement—no time. The discussion about whether or not there can be temporal duration without movement or change has been with us since then. The appropriate selection from Aristotle's writings may be found in Richard Gale's excellent reader *The Philosophy of Time* (Macmillan, London, 1968) which also contains articles by Plotinus and St. Augustine which criticize Aristotle's account. Further relevant readings on this topic may be found in J. J. C. Smart's reader *Problems of Space and Time* (Macmillan, New York, 1964) mentioned in previous chapters. Of particular relevance here would be the selections from the writings of St. Augustine, Isaac Newton, and Gottfried Leibniz. The solution to the frozen universe problem offered here is to the effect that the relationalist can retain his position, together with his being able to allow the possibility of a frozen universe, provided he is also willing to allow the possibility that the principle of determinism as commonly accepted is not the only possible deterministic principle. Section 4.3 was inserted primarily to forestall a possible objection to this solution, but also to criticize an argument form which is common in this branch of the literature. The quotation from Grünbaum on pages 75–6 is taken from his *Philosophical Problems of Space and Time* (Routledge & Kegan Paul, London, 1964), p. 197. Section 4.4 simply summarizes this chapter and prepares the reader for the following chapter.

Existence and the Present

5.1 WORDS AND CONCEPTS

Doubtless it is true that the way we conceive of many things is closely related to the way we are able to express ourselves. Some would go further and assert that the way we express our thoughts even affects the way we have of perceiving the world. This is not to say, of course, that our perceptions are limited to the extent that we possess abilities to describe our perceptions. It is just to say that different conceptual abilities and/or habits are liable to result in different ways of seeing the world.

Benjamin Lee Whorf, a linguist who has studied the Hopi language and culture, is one who believes the theory that the way we express our thoughts affects the way we think about the world about us. In his article 'An American Indian Model of the Universe', *International Journal of American Linguistics*, vol. 16 (1950), he claims that a Hopi has 'no general notion or intuition of time as a smooth flowing continuum . . .' and gives as evidence for this the fact that the Hopi language 'is seen to contain no words, grammatical forms, constructions, or expressions that refer directly to what we call "time" . . .'

Such a theory runs into the traditional 'other minds' problem as soon as one attempts to gather evidence in its support. For the only way one can know what a Hopi is thinking is to listen to what he says. But the theory seems plausible enough in the light of the personal experiences reported by those who have gained conceptual enrichment as the result of some educative process or other. To the layman, a well-kept lawn may seem simply like a vast array of closely clipped leaves of similar-looking grass. To the experienced gardener, that same lawn is 'obviously' a mixture of Kentucky blue-grass and African couch. It is not that the gardener has better vision than the layman. It is not even that the gardener notices details that the layman does not notice, though this of course will be so. It is that the gardener is capable of having different perceptual gestalts from those which the layman is capable of. But the matter would not end with *spatial* gestalts. It will also be the case, presumably, that whether or not we perceive a series of events as a single process, or as a sequence of disjoint processes, will depend upon how we think the events are causally related, which will in turn depend on how we *can* think of the events and their causal relationships, which will depend again on how we can formulate these thoughts. Indeed the very way we perceive and conceive of space and time may be biased by the manner in which we habitually describe the world.

But if two people perceive the world in different ways, does it follow that at least one of them is suffering from an illusion? No, for under normal circumstances no one would say that either the gardener or the layman was illuded when they gazed upon the well-mown lawn. Many people are wary of saying even that a person under the influence of an 'hallucenogenic' drug such as lysergic acid diethylamide (L.S.D.) is illuded. People who have experienced such a drug often wish to describe their experience as one of 'exquisite sensitivity to the environment', according to Sidney Cohen in an article in *Harper's Magazine* in 1965. Nevertheless, there are limits beyond which we do normally speak of an illusion having occurred. If a man had taken more than a normal dosage of L.S.D. and was seen to be lying on his back on the floor kicking his legs in the air, and if he later claimed to have seemed to be flying, we would believe that his perceptual apparatus had gone awry. And now, in a like manner, the question arises as to whether, in all sobriety, different ways of thinking of time and temporal succession could produce not just a peculiar way of perceiving the world but temporal perceptions that were actually illusory.

In particular, it is often claimed that our sense of time-flow—the feeling we have that events come out of the future, become present, and then recede into the past—is an illusion. Such a belief is common among the mystics of the Orient; but it is shared also by many physicists and philosophers of science. Thus on p. 132 of his book *Philosophy and Scientific Realism*, J. J. C. Smart maintains: 'Our notion of time as flowing, . . . , is an illusion which prevents us seeing the world as it really is.'

Smart claims that this illusion is largely maintained by the fact that we speak a tensed language. We use the so-called past tense in describing events which occur earlier than the time of making the description, the present tense in describing events which occur at a time simultaneous with our description of them, and the future tense in describing events which occur later than the description. This makes it seem that we are always describing events as falling into one of three categories: being in the past, being in the present, or being in the future. Thus we are able to say of some particular event that is now occurring, that it *was* in the future, it *is now* present, and it *will be* past.

In the remainder of this section, I shall argue that our ability to describe events as being in the past or present or future is not sufficient to give us the idea of a flow of time. The reason for this is that the situation as so far described could be deemed to be perfectly symmetrical between the past and the future. We use one sort of tense for the past, another for the present and yet another for the future. Even if we add the fact that what we mean by the past is anything *earlier than* the present, and what we mean by the future is anything *later* than the present, the situation could still be deemed to be symmetrical and static.

Consider a corresponding spatial analogy. Let us imagine that we use the words 'the planar' to refer to a plane parallel to the equatorial plane, and which passes through the speaker at the time of his uttering the words. Assume further, that we used the words 'the north' to refer to those events north of such a plane and 'the south' to refer to any events south of the plane. Further, let us assume that we modified our verbs depending on whether we were speaking of events which were to the north of us, or which were at the planar, or which were to the south of us, in the following way. If the event is to the north, we use the prefix 'nor' before the verb. If the event is at the planar, we use the prefix 'top'. If the event is to the south, we use the prefix 'sor'. Thus a speaker in Britain in August 1972 would say 'There soris a war in Indo-China.'

The correct grammar for a speaker in Australia would be 'There noris a war in Indo-China', while a person in Indo-China, or for that matter, in Thailand or Burma, would say 'There topis a war in Indo-China.' We could also say of any event that topis occurring, that it *soris* to the north, it *topis* at the planar, and it *noris* to the south. That is, speaking normally, south of here, the event is to the north, here the event is on this plane, and north of here, the event is to the south. Further, corresponding to the facts that the past is *earlier than* the present and the future is *later than* the present, there would be the facts that the south is *south of* the planar and the north is *north of* the planar.

In his *Philosophy and Scientific Realism*, Smart has claimed that if we were to rid our language of tenses and of the phrases 'in the future', 'at present', and 'in the past', we could defuse our tendency to think of time as flowing. The idea is that we use only one tense, or rather that we use a tenseless form of the verb, and that we use the descriptions 'earlier than this utterance', 'simultaneous with this utterance', and 'later than this utterance' to give the relative temporal position of the utterance with the event being described. In Smart's reduction 'this utterance' is meant to refer to the utterance at the time it is being produced. Thus 'There will be peace' is rendered 'There *is* peace later than this utterance', 'There is a war going on' is rendered 'There *is* a war going on simultaneous with this utterance', and 'There was a war' is rendered 'There *is* a war earlier than this utterance', where the italicized '*is*' is the tenseless 'is' of 'Two is an even number' or 'All ravens are black.' Confusion about time-flow, it is claimed, tends to arise due to the fact that our normal tensed manner of speaking fails to emphasize the *relational* nature of tenses, the relations being the temporal relations of 'earlier than', 'simultaneous with', and 'later than' between the events being described and the utterance which describes them. Likewise the phrases 'in the future', 'at present', and 'in the past' tend, we are told, to give us the impression that events can have a property of being in the future or being present or being in the past, whereas what is actually the case is that there are no such one-place properties of single events. What there are, are two-place relations between events, the relations of being earlier than, simultaneous with, or later than.

Now it might be true that the use of a tenseless language in which these relationships were made explicit would cure us of our time-flow illusion, if illusion it is. But how it would do this I do not know, for what is certain is that such an illusion could not be *explained* in terms of our tensed language and our ability to refer to the past, present, and future, particularly if our tensed language is regarded as being like its corresponding spatially tensed counter-part. For the corresponding spatially tensed language of the north, the planar, and the south illustrates the absolute symmetry of this mode of speech. Even if we were confused enough to think of 'being at planar' as an intrinsic property of events, rather than a relation between events and ourselves, there would be no reason to think of the planar moving northwards any more than there would be reason to think of it moving southwards. Yet we feel that the present encroaches on the future and that it does not encroach upon the past. We feel ourselves to be approaching death, not birth. Alternatively we sometimes think of the present as something static, past which events flow in a vast stream, as when we think '1972 will soon be past.' But there is no more reason to think of events to the north flowing past the planar into the south any more than there is to think of events to the south flowing past the planar to the north.

In future sections I shall from time to time wish to refer to two kinds of tensed language. If the job that the tenses do is, like that of the spatially tensed language mentioned in this section, merely to indicate one of two directions without any further metaphysical connotations such as that of 'flow' or anything else, I shall say that

the tenses are used *purely indexically*. Otherwise I shall say that the
tenses are not used purely indexically.

In section 5.2 it is shown how reductions to purely indexical
tensed discourse can eliminate reference to times and places, though
it is also argued in sections 5.3 and 5.4 that the tensed discourse of
ordinary language is not purely indexical.

5.2   THE DEBATE ON TENSE ELIMINATION

Considerable debate in recent years has centred on the attempt
by philosophers including Smart, Quine, Reichenbach, and Russell
to de-tense language, and rid our language of reference to the past,
present, and future. On the other hand, C. D. Broad, Richard Gale,
Arthur Prior, and others have thought it enormously important to
resist such attempts. What is interesting is that most of the
philosophers on both sides of the debate have thought that something would be lost in the de-tensing process. Those in favour looked
upon the loss as a loss of confusion—confusion which leads to
mythological beliefs if not downright paradox. Those against
regarded the loss as a loss of important conceptual abilities needed
for a complete description of the world.

The point is that the reductions were regarded as ones in which
there were conceptual change. Therefore these reductions cannot be
regarded as being analytic reductions in the sense in which I have
used that term. Hence if someone argues to the effect that reductions
such as Smart has proposed do not give a correct analysis of tenses,
and words such as 'past', 'present' and 'future', he or she would be
mistaken as to the type of the reduction being offered. For example,
in *An Examination of McTaggart's Philosophy*, vol. 2, Part I, Broad, in
discussing an attempt of Russell's, similar to Smart's, to be rid of
tenses, says that a speaker who says 'It is raining now' is not expressing a judgement about the utterance itself whereas someone who
says 'An occurrence of rain is spatio-temporally contiguous with this
utterance of mine' would (supposedly) be using the utterance to
express a judgement which he is making about the utterance itself.
He concludes that this is an objection to Russell's 'theory'. But it is
an objection only if Russell's 'theory' is an attempt at an analytic
reduction rather than an explicatory reduction.

A similar comment applies to a later remark of Broad's in this
same article, namely that 'the theory leaves altogether out of
account the transitory aspect of Time'. He says that qualitative
changes that take place in the course of one's experience are supposed
(by Russell) to be 'completely analysable into the fact that different
terms of this series' (of experiences) 'differ in quality, as different
segments of a variously coloured string differ in colour. But', he

proceeds, 'this leaves out the fact that at any moment a certain short segment of the series is marked out from all the rest by the quality of presentedness.'

However, it is just such confusing 'facts' as this that I presume Russell, Smart, and Quine would wish to leave out—for they would wish to reply, I presume, that certainly it is true that if any particular moment is present then all other moments are not present at that moment. But this is true of all moments, not just one. It is precisely because, with Broad's way of describing the situation, it is so easy to regard being present as a property of an event in the same way that being frightful or unexpected and so on are properties of events and that they regard this as a mistaken view, that they wish to *replace* Broad's way of describing temporal matters with one which does not lead to that sort of confusion.

Arthur Prior in his book *Past, Present and Future,* argues that being past, being present and being future are not characteristics. On page 18 he considers a conundrum raised by G. E. Moore—'How can an event have a characteristic at a time when it isn't?' Prior claims that 'is past' is not a predicate, it is but a quasi-predicate of a quasi-subject, namely an event. 'X's starting to be Y is past' just means, so he claims, 'It has been that X is starting to be Y', and the subject here is not 'X's starting to be Y' but X. He goes on:

> It is X which comes to have started to be Y, and it is of X that it comes to be always the case that it once started to be Y; the other entities are superfluous, and we see how to do without them, how to stop treating them as subjects, when we see how to stop treating their temporal qualifications ('past', etc.) as predicates, by rephrasings which replace them with propositional prefixed ('It has been that', etc.) analogous to negation.

What Prior is offering here is an analytic reduction of contexts containing 'is past', 'is present', and 'is future' to contexts which contain the correspondingly tensed sentences instead. Further he is prescribing this reduction to eliminate the difficulty raised by Moore. However, he objects to what he describes as the 'translation' of tensed sentences offered by Russell and Smart, on the same basis as does Broad—that it is not a satisfactory analysis of tensed sentences. As I have already suggested, this is not a satisfactory objection in itself if the reduction is deemed to be explicatory rather than analytic.

But what is a satisfactory objection to an explicatory reduction? It is this: that there are useful and/or interesting statements about the world that are expressible in the original mode of speech which are *not* expressible in the proposed new mode of speech. Thus the worthiness of the reduction of 'X is present' into 'X is simultaneous

with the production of this utterance' as an explication is not
impugned by the fact that in ordinary speech it is possible that some
event X is present even when no one is uttering some statement which
says that it is. For, as Gale correctly points out in his *Language of
Time*, Chapter X, we can never *say* that X is present without
producing an utterance. So a reduction of 'X is present' into 'X is
simultaneous with the production of this utterance' is on the surface
of it, satisfactory since 'X is present' is not *expressible* in the absence
of an utterance in which to express it. Gale, however, like Broad,
treats the reduction as an analytic reduction and concludes on that
basis that reduction is unsatisfactory. But this objection can be
discounted for the reasons given above.

I wish to show now that the satisfactoriness of this reduction is
only superficial in that unobjectionable objections can be brought
to bear against it. These objections will still be trifling, however,
since the reduction can easily be amended to overcome them.
However, a further paradox arises with the amended reduction.
Let us then return to Prior and Smart in order to examine, in greater
depth, Prior's objection to Smart's reduction.

On page 10 of *Past, Present and Future*, Prior discusses an attempt
by Smart ('The River of Time', *Mind*, October 1949) to deal with
the sentence:

(1) 'The beginning of the war was future, is present and will be
past' in terms of 'this utterance', 'earlier than', 'simultaneous with',
and 'later than'. Smart compares this sentence to (2) 'A boat was
upstream, is level and will be downstream'. This sentence he
reduces as follows:

Occasions on which the boat is upstream are earlier than this
utterance; the occasion on which it is level is simultaneous with this
utterance, and the occasions on which it is downstream are later
than this utterance.

Smart then claims, quite correctly, that we cannot reduce sentence
(1) in the same way. If we try, he says, we get:

(3) 'The beginning of the war is later than some utterance earlier
than this one, is simultaneous with this utterance, and is earlier
than some utterance later than this one.' Wherein 'this utterance'
occurs once only—not three times.

Prior points out that (1) is the equivalent of:

(4) 'The war was going to begin, is now beginning and will have
begun' which is not a conjunction of simply tensed sentences as (2)
is, but is rather a conjunction of sentences using more complicated
tensing. He goes on to state that Smart's analysis of tensed utterances
'is quite implausible even when the tenses are simple. . . . But when
it is applied to tenses such as the future perfect, it becomes downright

fantastic.' He later goes on 'How are we to analyse, for example, "Eventually all speech will have come to an end"? What Smart's recipe would give us is "The end of all utterances is earlier than some utterance later than this one", which translates something empirically possible into a self-contradiction.' Prior concludes that 'the real moral of Smart's paper is that the Russellian analysis of tenses breaks down . . . , as soon as we remember that there is such a tense as the future perfect.'

In 'The River of Time' and also in *Philosophy and Scientific Realism* Smart, too, seems to be of the opinion that sentences such as (1) cannot be dealt with in terms of 'earlier than', 'simultaneous with', 'later than', and 'this utterance'. On page 134 of *Philosophy and Scientific Realism* he says that there is one thing 'that we cannot say in our tenseless language. We cannot translate a sentence of the form "this event was future, is present and will be past".' He goes on to claim that this is a step in the right direction, because we don't wish to be able to talk of events changing in respect of pastness, presentness, and futurity.

However, a reduction of 'This event was future, is present and will be past' is possible in terms of 'earlier than', 'simultaneous with', 'later than', 'this utterance', and the tenseless '*is*', so long as one is willing to countenance the existence of times as well. Here it is:

(5) This event *is* later than a time which is earlier than this utterance, it *is* simultaneous with this utterance, and it *is* earlier than a time which *is* later than this utterance.

Likewise 'Eventually all speech will have come to an end' may be rendered:

'All utterances *are* earlier than a time which *is* later than this utterance.'

What Prior did, in effect, was to reject a reduction, because a 'recipe' for the reduction had not been found which could be applied to all the contexts to be reduced. But to ask for 'recipes' for a reduction that will work throughout the length and breadth of the complicated syntax of a natural language is a tall order. As any linguist will grant, there is usually an exception to any grammatical rule in a natural language. The best someone can do in proposing a reduction is to give examples of a few paradigm cases of its application and hope to be able to invent solutions to would-be counter-examples as they arise.

One thing that can be said about (5), however, is this. If we are willing to countenance the existence of times, we might as well dispense with references to 'this utterance' in (5), in favour of 'this time' or, less ambiguously, 'now'. This would render (5) into:

(6) This event *is* later than a time which *is* earlier than now,

it *is* now, and it *is* earlier than a time which *is* later than now.

The introduction of times does not automatically introduce a flowing of these times from the future through the present into the past, for the corresponding (7) for places has no such implications. (7) This event is to the north of a place which is to the south of the planar, it is at the planar, and it is to the south of a place which is to the north of the planar.

However, if we have to introduce times to rid ourselves of tenses, then the same sort of difficulty arises for us as it did for Melissus, Leibniz, and Descartes with respect to places. These philosophers thought that there could be no place where there was a vacuum, i.e. where there was nothing—for there would always be a *place* there. Likewise if we take the existence of times seriously, it seems that we must assert that there can be no time when there is nothing, since at any time there will at least be a time at that time. But if we were so bound, then we would be bound *by our language* into the assertion of what seems to be a contingent truth (or falsehood) about the way the world happens to be. Superficially, at least, it would appear that our language would restrict us from the assertion of certain cosmological theories.

Alternatively, since tenses used purely indexically do not seem to introduce any metaphysical difficulties of their own (or at least none that we have so far discovered), perhaps we could employ them as Prior did, in the reduction of some statements involving times.

Thus the paradoxical:

'There is a time when there is nothing',

becomes:

'It either was the case, is the case or will be the case that there is nothing.'

One may object that a mention of an actual date presents a difficulty. For if it was the case that Elizabeth I died in A.D. 1603, then surely it would follow that there is a *time* at which Elizabeth I died. Indeed it would. But there would be no need to render the implication in this way, for we could instead render it thus: 'It was the case that Elizabeth I dies'. To put the matter another way, instead of regarding 'A.D. 1603' as the *name* of a *time* when Elizabeth I died, we now regard A.D. 1603 as *when* Elizabeth I died.

In a similar way, we could rid our spatial descriptions of references to points, places, and positions in order to escape Melissus's spatial paradox. If we ever felt the need to say 'There is a place where . . .' we could instead move to the spatially tensed talk of 'soris', 'topis', and 'noris' and say instead 'Either it soris the case that . . . , or it topis the case that . . . , or it noris the case that . . .'

More or less redundant expressions such as:
'There is a point five inches from the end of this pencil.'
and
'There was a time five minutes ago',
could be replaced by the equally redundant:
'Either something or nothing is five inches from the end of this pencil'
and
'There was either something or nothing five minutes ago.'

To sum up the results of this section, I have tried to show that certain objections to attempts to de-tense language are misconceived. Nevertheless, as it was argued in section 5.1, if tenses are regarded as being purely indexical, the need for these de-tensing reductions seems to vanish. Finally it was suggested that certain reductions in the reverse direction to purely indexical tensed language may be of some use in the elimination of philosophico-linguistic problems concerned with the existence of times and places.

5.3 EVENTS, TENSE, AND EXISTENCE

In section 5.1 it was argued that our ideas of time-flow cannot arise directly from tensed language if tenses are purely indexical, and this was done by comparing normal temporally tensed language with an artificial language using spatial tenses.

Yet the present seems real to us in a way that the past and the future do not.

'What happened to the blackberry pie?' a child asks anxiously.
'It's all gone,' says his brother.
How sad. But,
'No it's not,' says his sister,
'There's some left in the pantry.'

Oh, joy. For the child, reality is what exists *now*, even if it does not exist *here*. It is not that the child just wants to be contiguous with the pie somewhere and sometime. For he would hardly have felt joy had his sister merely reminded him that he had already eaten some of the pie. That event is already finished, gone, and is no longer a reality.

Again, the memory of a bereavement engenders less sadness within us as time passes and, likewise, the knowledge that such a bereavement will some day occur is less saddening the further into the future we believe that bereavement to be, but the grief of a bereavement is not mitigated by spatial distance from the death of the beloved.

Is there, then, some important ontological difference between what *noris*, *topis*, and *soris* on the one hand, and what will be and was on the other. If temporal tenses are purely indexical, then, as we have argued the difference—if there was one—would *not* be a matter of logic. It would be a matter of *contingent* fact. That is, we could deny that there is such a difference without contradicting ourselves. Let us for the moment assume that there is such a difference. Then the theory is, presumably, that in some sense of 'exists' whatever noris, topis, or soris exists; and whatever will be or was, does not exist. In saying this, it is important to note that it would be useless to have 'exists' simply *meaning* 'noris or topis or soris', for 'noris or topis or soris' is simply a long-winded way of saying 'is' or 'is now' or 'is present'. This being the case, the proposedly informative:

'Present events exist; all others do not exist'

would analytically reduce to the tautological

'Present events are present; all others are not present.'

That is, the predicate 'exists' would not give us the ability to make a distinction that we did not already possess. For the same reason 'exists' must be regarded as being tenseless, for if it is regarded as being in the present tense, then once again it is tautological and completely uninformative to remark that an event exists only if it is present.

Let us then write 'Exists' rather than 'exists' to indicate this tenselessness.

Our theory now becomes: 'All present events Exist but no other events Exist.'

But can 'Exists' be allowed to be tenseless in the sense that it means 'either existed, exists, or will exist'? If so we are still left with the problem of interpreting the tensed verbs 'existed', 'exists', and 'will exist'. We are also faced, by the way, with a slight stretching of English grammar, for, although it is grammatical to say that numbers exist and that objects and organizations existed or will exist, it is slightly ungrammatical to say that events exist or existed. More normal usage would be to say that the event actually happened, or is really happening, or will really occur. But we shall ignore these niceties for the moment and use the verb 'to exist' for events as well as other things. What is worse is that we are faced with the following sort of deductions:

Henry VIII's birth existed.

∴ Henry VIII's birth either existed, exists, or will exist.

∴ Henry VIII's birth Exists.

Again

Charles III's coronation will exist.

∴ Charles III's coronation either existed, exists, or will exist.

∴ Charles III's coronation Exists.

Add to these conclusions the assumption we have been considering, namely that an event Exists if and only if it is present, and we obtain:

Henry VIII's birth is present,

and

Charles III's coronation is present.

—both of which are falsehoods.

If we were to stick to our assumption then, and also the interpretation of 'Exists' as 'either existed, exists, or will exist', then we would have to deny that:

Henry VIII's birth existed,

and also that:

Charles III's coronation will exist.

But the first of these statements is true, and the second is at least highly probable.

In any case, if 'existed' is not a predicate which would truly apply to Henry VIII's birth, we are still left with the question of what it *does* mean. And the same question also remains for the present-tensed 'exists' and the future-tensed 'will exist'. And, while these are left without any clear meaning, we have not succeeded in making explicit the meaning of the tenseless 'Exists' if the only possibility is to define it as 'existed, exists, or will exist'.

Perhaps the tenseless 'Exists' should be regarded as tenseless in much the same way as the 'is' in 'two plus two is equal to four' is tenseless. For similarly in this case it is rather bizarre to interpret the 'is' as 'either was, is, or will be'. To say that two plus two either was, is, or will be four would be to invite the possibility that there were times when two plus two was not equal to four, or that there will be times when two plus two will not be equal to four or even that two plus two is not equal to four now. If anything, two plus two was always equal to four, is equal to four, and always will be equal to four, which implies the weaker statement that 'Either two plus two was equal to four, is equal to four, or will be equal to four', but not vice versa. We might consider, then, the interpretation of 'Exists' as 'always existed, exists and always will exist', ignoring, for the moment, that we still don't know what sense to give to 'existed', 'exists', and 'will exist'.

Thus we shall have:

(1) An event Exists if and only if the event always has existed, exists, and always will exist.

Now I have just said that we don't know what the present-tensed 'exists' amounts to, but it is at least known to be present-tensed. It would seem reasonable then to assume:

(2) If an event exists then the event is present.

Further, it seems reasonable to add the further assumptions:

(3) An event has always existed if and only if it has always been the case that the event exists
and
(4) An event will always exist if and only if it will always be the case that the event exists.

What happens if we add to these our original assumption, namely:
(5) 'An event Exists if and only if it is present.'?

It follows from these assumptions that an event Exists if and only if it has always been the case, it is the case, and always will be the case that the event is present.

That is, an event Exists if and only if it is of eternal duration. What is even less palatable, perhaps, is that it also follows that an event is present if and only if it always has been and always will be present. Thus if I am at present writing this book, it would follow not only that I always have been writing this book, but also that I always will be doing so, and this not only would be unendurable, but is, I am glad to say, false.

Well, let us start again. The object of the exercise was to cook up a possible but non-trivial ontological difference between what is on the one hand and what was or will be on the other. Let us first assume that there is something, which is The Present. Let us further assume that The Present does as a matter of contingent fact Exist and that nothing else Exists. Let us now *stipulate*, in order to give 'The Present' some meaning, the following:
(6) If something is The Present, then it always was, and always will be The Present.
(7) An event is present if and only if it coincides with The Present.
(8) An event exists if and only if it coincides with something that Exists.

One more assumption is needed; one which is analytically true. It is:
(9) Everything coincides with itself.

It follows from all these assumptions and stipulations that:

   (i) The Present always was present, is present, and always will be present.

  (ii) The Present always existed, exists, and always will exist.

 (iii) Any event which coincides with The Present exists and is present; but if it is not The Present itself, then either it was not always the case that it existed or it will not always be the case that it exists.

 (iv) An event which does not coincide with The Present is not present, nor does it exist.

Why is it, then, that my writing this book is present? According to this theory, it is because my writing this book coincides with The Present. And it is this coincidence which gives it its existence. Also when I have finished writing the book, my writing of the book will no longer coincide with The Present, so my writing this book will no longer be present and my writing this book will no longer exist. Thus events to which one can give a date, such as the writing of this book, or the birth of Henry VIII, or the coronation of Charles III, exist at some time, but do not Exist. Let us call such events *datable events*. Datable events, then, borrow their existence, as it were, from The Present by being coincident with The Present. Only The Present exists at all times and hence only The Present Exists.

Of course, one could have made corresponding stipulations about The Planar as well, using spatially tensed constructions. But the assumption that only The Present Exists is a contingent assumption. It is not part of what is meant by 'The Present'. It is part of this theory, then, that The Planar, unlike The Present, does not Exist.

McTaggart, in his book *The Nature of Existence*, tried to differentiate between two series of events which he called the *A*-series and the *B*-series. Some of the properties McTaggart ascribes to the *B*-series seem to fit our datable events, but some of the properties of the *A*-series also fit our datable events. For example the *B*-series events are said to be such that if one event, x, in the *B*-series is later than some other event, y, then it will always be so. Thus the birth of Henry VIII and the coronation of Charles III (assuming that it will occur) certainly fit these *B*-series descriptions.

The characteristics of *B*-series events are, however, said to be permanent, while the characteristics of *A*-series events are not. Characteristics of *A*-series events are being either Past, Present, or Future. But Charles III's coronation has now the impermanent characteristic of being future and, although the future permanency of the pastness of Henry VIII's birth seems now to be guaranteed, it once was future, became present, and is now past. So this determinant of what McTaggart called the *A*-series, also seems to apply to datable events.

On the other hand in a footnote he seems to think we can consider the *B*-series as 'sliding along' a fixed *A*-series, in which case 'time presents itself as a movement from future to past'; alternatively we are allowed to consider the *A*-series as sliding along a fixed *B*-series, in which case 'time presents itself as a movement from earlier to later'. This makes the *A*-series look like a series of events, one of which is the present, another of which would be the event which is always one unit of time into the future, another would always be two units of time into the future and so on, and others would be various temporal distances into the past. Under this interpretation of the *A*-series and the *B*-series, *B*-series events receive their transitory characteristics of being past, present, or future, by virtue of being coincident momentarily with an *A*-series 'event' which is *permanently* past, present, or future as the case may be. *A*-series events, like *B*-series events, would retain their relationships of being earlier than, simultaneous with, or later than one another, but unlike the *B*-series events, they would also retain their properties of pastness, presentness, or futurity. Also, unlike the *B*-series events, all other characteristics of *A*-series events undergo change. The present, for example, is no longer to be truly described as encompassing the birth of Henry VIII, though it once did. The *B*-series event, however, which is the birth of Henry VIII, always has been and always will be the birth of Henry VIII.

What I am now describing as *B*-series events, then, is what I have called datable events. These events never change their spatial characteristics nor their temporal relationships to other *B*-events. Their only change is in their temporal relationships to *A*-events and thus in their characteristics of pastness, presentness, and futurity. *A*-events, on the other hand, may change their spatial characteristics continuously, as well as their temporal relationships to *B*-events, though they never change their temporal relationships to other *A*-events and thus never change their pastness, presentness, or futurity.

What I have referred to as 'the present' in this description of the *A*-series will of course be The Present. For The Present is that which never changes its state of being present. This probably takes us a fair way from what McTaggart meant by *A*-series and *B*-series events. But it is not at all clear to me, at least, what he meant anyway. Certainly my concern here is *not* (as it was for McTaggart) whether the *A*-series or the *B*-series is real. What I have been considering is a theory which claims that one and only one event in the *A*-series, namely The Present, Exists while all other events in the *A*-series do not Exist and do never exist. Also events in the *B*-series, in this theory, exist only for finite durations. They exist when they coincide with The Present, but none of them Exist.

To sum up so far, this section began with the question of whether or not there was an ontological difference between events that are truly described in the present tense, as opposed to the past and future tenses. It was argued that if these tenses were purely indexical, then this difference (if there is such a difference) would not be a matter of logic, but rather a matter of contingent fact. Within a purely indexical interpretation of the tenses, a contingent theory involving an entity called 'The Present' was then developed, according to which present events derive their existence by coincidence with The Present, which in turn Exists in a tenseless sense. The object of developing this theory is to use it by way of comparison in subsequent sections to bring to the surface some of the metaphysical assumptions that lie behind the common use of tensed expressions.

5.4 MOORE'S PROBLEM AND THE PRESENT

It is time now to have a closer look at Moore's problem which we mentioned in passing in the section 5.2, namely 'How can an event have a characteristic at a time at which it isn't?' Now if we swallow the above story about The Present, then we must also believe that it is perfectly reasonable to refer to and describe, as being past or future, events which neither exist nor Exist. Prior's way of dealing with Moore's problem, it will be recalled, is to cease to refer to events and hence to eliminate descriptions of events, and to refer instead to objects and to use tensed language in the description of these objects. If one adopted Prior's prescriptions in this matter, then the above theory could not be formulated, for the formulation depends on the predicate 'is present' which, in the sense in which it is used here, is one of the predicates that Prior's reformulations eliminate. In this section it will be argued that Prior's suggested reformulations do not in themselves eliminate *all* problems of the sort raised by Moore—as Prior himself realizes; for the same kind of problems arise for objects as for events. Also it will be argued that the fact that Moore's problem is a difficulty for both Moore and Prior, indicates that these philosophers are not using temporal tenses in a purely indexical way, but that in addition, their use of tenses also presupposes the sort of beliefs made overt by the theory of The Present—for, so it will be argued, Moore's problem does not necessarily arise if tenses are used in a purely indexical way.

When Prior is dealing with Moore's problem he says that 'X's starting to be Y is past' just means 'It has been that X is starting to be Y', and the subject here, he claims, is not 'X's starting to be Y', but X. He goes on to remark that there is no need to think of 'X's starting to be Y', as 'momentarily doing something called "being present" and then doing something else called "being past" for much longer . . . It is X which comes to have started to be Y . . .; the other entities are superfluous.'

A method based on this suggestion, for ridding our discourse of what seemed to be reference to times or temporal instants, has already been suggested in section 5.2 for dealing with the problem of whether or not there can be times when there is nothing. A similar method was suggested also, using spatial tenses, for dealing with Melissus's paradox to the effect that there could be no vacuums. But events are not times any more than objects are places. To dispense with events, so I shall argue, is to go too far.

Events, physical objects, and processes all have this in common, that they are usually thought of as having length, breadth, thickness, and temporal duration. It is true that physicists and philosophers sometimes use the word 'event' to mean an instant or moment of time—something of zero temporal duration or an infinitesimal temporal duration—and in this sense the word is often used to mean something which covers all of space, but often it is also used to mean something which covers only a finite amount of space. But when people say that the third event of the athletics meeting is the 1500 metres race or that the civil war during the reign of Charles II was a decisive event in the history of Britain, they are referring to *something which happens* within finite spatio-temporal boundaries—and not just the space–time within those boundaries.

The difference between events in this sense and physical objects and processes seems rather nebulous. Events are usually considered to be of small temporal duration whereas physical objects and processes are usually considered to be of relatively long temporal duration. Physical objects do not change very much or, if they do, the change is not rapid. Processes are of interest because of the changes involved. It is change which marks out a process. A physical object is usually distinguished by more or less static properties. The distinctions just made are not definite for they depend on the attributive adjectives 'long', 'short', 'rapid', and 'more or less static'; and just as big mice can be small animals, so what is long in one context can be short in another, and what is rapid with respect to one set of phenomena can be slow in comparison with something else. Thus a conference of diplomats may be regarded by one of the participants as a long-drawn-out process, and by a historian fifty years later as an important event. A cloud above a mountain divide is considered by someone in an aesthetic mood to be an object reflecting the colours of the setting sun and by someone in a meteorological frame of mind as a continuous process of condensation and subsequent re-evaporation of the water vapour in the wind passing over the mountain. There are differences in the way we refer to the temporal boundaries of events, objects, and processes. Thus events such as, for example, the shearers' strike of 1891 are said to happen at some specified time or during some specified duration of time; processes are said to begin and end, physical objects to come into existence or be created, and to be destroyed or pass out of existence. People's lives come to an end, but people die and their bodies are destroyed or decay. These 'differences' in the ways physical objects, processes, and events begin and end their existence in time seem grammatical rather than semantic. Yet there are, I think, some semantic differences. Events and processes are happenings—changes of state. Physical objects are not. Changes of state can occur within a physical object, or on the surface of a physical object or between physical objects as in a collision. A state is a property instantiated by a physical object or a set of physical objects. Events, in the sense that the First World War is an event, and processes are changes in such states.

Thus, to hark back to a previous example, the aesthete's cloud is not to be identified with the meteorologist's condensations and evaporations. The meteorologist's processes are changes in state of the mixture of air and water flowing over the mountain. The cloud is not a change of state. It is something which *has* states such as reflecting sunlight, being electrically charged, and so on. Yet both the cloud and the meteorologist's processes lie within the same spatio-temporal boundaries.

Events, in the sense used here, are not *just* slices of time, finite or infinitesmal. Neither are they *just* chunks of space–time. They are physical goings-on. One can have a chunk of space–time without anything going on within it. That would not be an event in this sense. Also, events can have properties other than being past, present, or future. And although one can, in ordinary discourse, apply these predicates or quasi-predicates to events, one can apply other predicates to them also, and that in contexts where no particular physical object is referred to. One might agree that no events in this sense could occur unless there were physical objects which partook in the events. But that is not to say that the descriptions of such events are descriptions of the corresponding physical objects. When we say that the time for (i.e. the duration of) the relay race was fifty-five seconds, we are not describing some physical object, let alone the runners, as being of fifty-five seconds' duration. Likewise if it is said that the race was exciting to a spectator, it is not implied that the runners were exciting to the spectator, though, of course, that may have been the case as well.

But we do not have to dispense with events in order to dispense with 'past', 'present', and 'future'. Instead of saying that the third event on the programme is in the past, we could always say instead, as we probably would, that the third event on the programme is over, or has already occurred.

But even if, as Prior claims, all descriptions involving the word 'event' were analytically reducible to tensed descriptions of physical objects, this would not eliminate the sort of problems raised by Moore. For the question:

'How can an event have a characteristic at a time at which it isn't?'

can be replaced by:

'How can an object have a characteristic when it isn't?'

One might ask 'Can it?' and the answer is yes, at least if we take the following locutions seriously:

'Ethelred is dead.'

'The man who can beat Bobby Fischer at chess is not yet born.'

Perhaps such predicates as 'is dead' and 'is not yet born', which seem to describe objects which do not exist, can also be eliminated with appropriate analytic reductions. Perhaps we could translate 'Ethelred is dead', for example, by 'Ethelred died'.

Let us assume that such reductions could be carried through. There would still remain the predicate or quasi-predicate 'exists'. How can we say of things—be they events or objects—that they do not exist? Thus King Harold and the Battle of Hastings existed, but they do not now exist. It is no use trying to translate this as 'King Harold and the Battle of Hastings existed' for it is consistent with this that they still exist.

Some would say that the trouble here is that we are treating 'exists' as a predicate, which it isn't. To use an example of Moore's, 'some tame tigers growl' means:

'There are some tame tigers which growl.'

If we used the same technique for 'some tame tigers exist', we would come up with:

'There are some tame tigers which exist'

which is queer. The clause 'which exist' does not describe the tame tigers. When we compare the cases:

'Some tame tigers do not growl'

and

'Some tame tigers do not exist',

the absurdity of treating 'exist' as a predicate seems clear. The first of these may be rendered:

'There are some tame tigers which do not growl',

but the second, using the same technique becomes:

'There are some tame tigers which do not exist',

which seems absurd.

Very well. So, to avoid such confusions, we could, instead of using such expressions as 'Tame tigers do not exist' use instead 'There are no tame tigers' or 'No tigers are tame'. But what should we do if the verb 'to exist' occurs in the past or future tenses? How are we to deal with 'Dinosaurs existed but they no longer exist'? I suppose what would come fairly naturally is 'There were dinosaurs but there aren't any dinosaurs any longer.' So what is the difficulty? One of the difficulties in giving a thorough-going reduction and a logic for all such cases, rather than particular examples, is the immensity of the task. The task is to produce a consistent way of describing events and objects in the past, present, and future that does not violate the dictum presupposed by Moore's problem, namely, that there are descriptions that are true of a thing *only* when the descriptions apply to the time that the thing existed, exists, or will exist. What is to be eliminated are descriptions of a thing that are true of that thing only at times when it doesn't exist. That is why 'Ethelred is dead' was a problem. It is a true description of Ethelred only when Ethelred no longer exists. The analytic reduction of such expressions to expressions which do not violate Moore's presupposition is a task which Prior and others have set themselves. Thus, for Prior, 'I am thinner than my great-grand-father' is a problem, for such a statement seems to describe my great-grandfather who no longer exists, and a similar problem arises out of comparisons between any two objects or events that do not co-exist. The expression of some such relationships can be handled easily. For example, if *x* was the grandfather of *y*, *y* may never have existed at a time when *x* did. But '*x* is the grandfather of *y*' can be rendered '*x* was the father of someone who is (or was) the parent of *y*'. However, '*x* was thinner than *y* is', where *x* and *y* do not co-exist, cannot be so easily dealt with. Prior suggests that we could overcome this by expressing the proposition in terms of girths. The appropriate analysis would be:

'For some girths *f* and *g*, *x*'s girth was *f* and *y*'s girth is *g* and *g* is always less than *f*.'

But it is rather strange that we have to invent timeless entities such as girths just so that *x* can be simultaneous at one time with a girth that always co-exists with another girth which in turn co-exists with *y* at some other time.

The fact that Prior and other logicians interested in tense logic feel the need to go to such lengths shows that temporal tenses so conceived are not purely indexical. For no one will feel ontologically upset at the thought of my being thinner than my brother, even if they think that my brother did not live on the same latitude as myself and never did. No one believes that something is somehow ontologically inferior because it has never been at any place where he has been or is or will be. To be sure, there are many properties, for example being red, or being of such and such a mass, or being of such and such a shape, which can be instantiated only at the places occupied by the objects that have those properties; but two objects do not have to have occupied the same place in order to bear relationships to one another. If what I'm pointing at is to the north of what you are pointing at, it is true that both indicated objects are not co-topexisting, but we don't for that reason insist that, if they are to bear some relationship to one another, there has to be some third object which overlaps spatially with each of the first two. If we did insist that there had to be such an overlap of objects for spatial relationships to be instantiated between objects, and if there were no physical objects around to do the job, could we, as Prior did with the case of '*x* is thinner than *y*', introduce some abstract entity such as a girth which topexisted everywhere? But what girth should one choose? A girth of twenty-six inches or thirty-four inches? Perhaps, to be fair, we should include all possible girths. The absurdity, of course, is that abstract entities such as girths, heights, and for that matter durations do not exist in space and time at all. They are neither physical objects, processes, or events. One can say, of course, should one choose to do so, that a girth of thirty inches is instantiated in such and such a tree or person and hence that instantiations of girths exist in space and time. But instantiations of girths are not girths. If one insists that objects, events, and processes can bear relationships to one another only if either they temporally overlap or if there is a succession of temporally overlapping objects between them and if one also insists that the spatial and temporal distribution of such events, objects, and processes is a contingent matter and that there are many relationships between such things that depend only in a contingent way if at all upon these distributions, then what one needs is an entity such that the entity is not a physical object, event, or process, and such that the entity is omnipresent.

When one adds to this the consideration that objects endure in time, and that an object's identity throughout the time of its endurance is guaranteed only by the instantiation of time-spanning relationships between earlier stages of the object and later stages of the object, then the omnipresent entity begins to look a little like the god that Aquinas needed to explain the continued existence of objects. It begins to look also very much like The Present, for in the theory of The Present, it will be recalled, something existed if and only if it coincided with The Present. Like Aquinas' god the existence of The Present was dependent on no other thing.

Now there is another though related way in which temporal tenses in their common use are not purely indexical. When we introduced the notion of the planar, and along with it our spatial tenses we did so with an arbitrariness which is lacking in the notion of the present and the use of temporal tenses. The surfaces of cotopexistence were chosen to be planes parallel to the plane of the equator. But one might just as well have chosen them to be concentric spheres with their centre, the centre of gravity of the earth, for example. Thus given an appropriate choice of a set of surfaces of co-topexistence, any two objects or events at all, could be said to co-topexist. Since one has this freedom in the choice of surfaces of co-topexistence, there is no point in postulating the existence of some omnipresent entity in order to guarantee the possibility of the instantiation of relationships between spatially separated objects. On the other hand, if there is a need to postulate the existence of some such entity in order to guarantee the possibility of the instantiation of relationships between *temporally* separated objects, then there is no freedom in the choice of the 'surfaces' of co-existence. That is, there would be no freedom to *choose* which events are simultaneous with which. Conversely if there were such freedom of choice, then we would also be free to choose whether or not to couch the description of some object in the past tense, future tense, or present tense. But normal usage of temporal tenses, so I would claim, presupposes that there is no such freedom. For the normal use of tenses to be operative, simultaneity must be deemed to be absolute. The planar can be of our own choosing, but there is only one Present.

The theory of The Present, then, combines in one package three aspects of tensed discourse. Firstly, it provides a framework for the indexical character of tenses common to all tensed discourse. Secondly it provides the continuing entity needed by the sort of tensed discourse typified in Prior's usage, for the instantiation of relationships between objects which do not co-exist, and thirdly, it provides a basis for the absoluteness of simultaneity presupposed by normal tensed discourse. I have not shown that it follows from undeniable premises that the theory of The Present is presupposed in the standard use of tensed discourse. But the theory that the theory of The Present is presupposed in the standard use of tensed discourse is a *contingent* theory about *all* standard usages of tensed discourse; and one cannot offer proofs for contingent theories about all things of any class unless it is possible to inspect every individual member of that class. One can but offer evidence in favour of such theories.

To summarize, this section dealt with Prior's treatment of a problem raised by Moore, namely 'How can an event have a characteristic at a time at which it isn't?'

Prior's solution, it was said, was to provide a reduction of expressions involving 'events' to expressions which referred only to objects. It was then argued that this solution was unsatisfactory in two ways:

(*a*) the sacrifice of events from our ontology needlessly limited the ways in which we could describe the world;

(b) Moore's problem remained with respect to objects anyway, even if it was recognized that 'exist' was not a predicate.

It was claimed that the fact that the difficulty prevailed showed that Moore and Prior were not using tenses in a purely indexical way. It was then argued that the Theory of The Present makes overt the ontological presuppositions in Prior's use of tenses as well as the presupposition that simultaneity is absolute which is inherent in standard tensed discourse.

5.5 SIMULTANEITY, THE THEORY OF THE PRESENT, AND MAXWELL'S THEORY OF ELECTROMAGNETIC RADIATION

The theory of The Present is contingent in so far as there may or may not be a thing that is The Present. If there is such a thing, then if there are objects, events, and processes, then these objects, events, and processes form a class of things which co-exist with one another by virtue of their being coincident with The Present. Such a class is, of course, supposed to be a class of simultaneous things. The simultaneity of the members of such a class is guaranteed by their coincidence with The Present and no other fact matters. Thus the simultaneity of events, in this theory, is absolute. It in no way depends on the choice of a frame of reference. Conversely if the simultaneity of events is relative to a frame of reference rather than absolute, then there is no such thing as The Present.

The case against absolute simultaneity derives from a consideration that was mentioned previously in section 2.6, namely that Galilean kinematics, Maxwell's theory of electromagnetic radiation, and the Restricted Principle of Relativity are inconsistent with one another. In particular, if one assumes Maxwell's theory to be true and also the restricted principle of relativity to be true, then it follows that there can be something, namely electromagnetic radiation, which would have a relative speed, c_0, with respect to any other things whatsoever, regardless of whether or not those things were in motion relative to each other. This, it was said, violated Galilean kinematics. But what is of importance here is that in particular, it violates the theory that simultaneity is absolute. It is the purpose of this section to show why this is the case and hence to show how Maxwell's theory and the Principle of Relativity, taken together, are inconsistent with the theory of The Present. Consider Fig. 19.

Let OO' be the chosen origin for the implementation of any spatial measuring system obeying the axioms of measurement referred to in section 3.6. Distances in the directions OX and $O\bar{X}$ represent spatial distances. Distances in the direction OO' represent durations

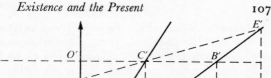

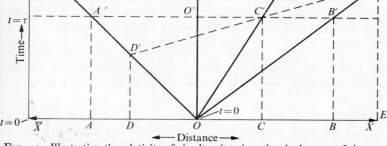

Fɪɢ. 19. *Illustrating the relativity of simultaneity given the absoluteness of the speed of electromagnetic radiation*

of time. A clock is started from zero at the same time as some electromagnetic radiation is flashed out from O in each of two opposite directions. The positions of the leading edges of the radiation at various times thereafter are represented by the lines OA' and OB'. At some time $t = \tau$, say, the left-hand ray reaches the position represented by the line, AA', and at the same time τ the right-hand ray reaches the position BB'. Since, according to Maxwell, both rays travel at the same speed, the distance OA will be equal to the distance OB. Meanwhile, a traveller, moving with respect to OO' at half the speed of the electromagnetic rays, has set off from O at the same time as the electromagnetic rays left O. His course is plotted by the line OC'. He reaches the position indicated by the line CC' at the *same* time τ that the rays reach AA' and BB'. The events represented by A', O', C', B' are then simultaneous. If the spatial measurement system obeys the axioms of section 3.6, it should be the case that the distance from C' to A' is greater than the distance from C' to B'.

Now consider the same situation with respect to another choice of origin of the *same* spatial measurement system. For origin, let us choose instead the position determined by the position of the traveller OC'. Given the truth of the Restricted Principle of Relativity, then, if Maxwell's theory is true, it should apply to this new frame of reference also, so that at any time the two rays should be equidistant from this new origin. Now if simultaneity is absolute, that is, if the same sets of events are simultaneous regardless of our choice of reference systems, then, if A', O', C', and B' were all simultaneous events when OO' was the spatial origin, so are these events simultaneous when OC' is chosen as the spatial origin. Hence for our traveller as for anyone else, there is no time-lapse at all between these events. What distance there is between them is purely spatial. Given then, the principle of independence of reference (see section 3.6, page 54), both reference-frames should yield the same distances from C' to A' and from C' to B'. But this would imply that for the 'travelling' frame also, the distance from C' to A' is greater than the distance from C' to B'. But as we have seen, on the assumption that Maxwell's theory is true and the Restricted Principles of Relativity also holds, the distance from C' to A' would have to be *equal* to the distance from C' to B' in the new frame of reference.

So, if we

(a) retain what we mean by distance (and so retain the axioms of spatial measurement)

(b) assume that electromagnetic radiation has the same velocity with respect to any reference-frame, and

(c) assume that the restricted Principle of Relativity is true,

there is only one assumption left to reject; and that is the assumption that with respect to our travelling reference-frame, the events A', O', B', and C' are simultaneous. Some other set of events, represented by the line $D'C'E'$, say, would have to be simultaneous with respect to the 'travelling' frame of reference for the assumptions (a), (b), and (c) to hold true. Thus if one assumes (a) and (c), then any evidence in favour of (b) would be evidence against the proposition that simultaneity is absolute and hence be evidence against the theory of The Present. Let us assume also that such evidence is forthcoming. Then one should conclude in all consistency that simultaneity, as a two-place relation between events, is uninstantiated in this universe. Also one should conclude that if one is to retain some notion of simultaneity, that notion should be a three-place relation between two events and a reference-frame. Einstein's Special Theory of Relativity may be regarded as the theory within which these consequences are accepted, and within which, also, a particular notion of three-place simultaneity is made explicit.

The questions which now arise, of course, are 'Are Maxwell's laws of electromagnetic radiation true? What evidence for or against such laws could one provide experimentally? Does Einstein's explication of simultaneity guarantee *a priori*, that is, independent of experimental observations, that electromagnetic radiation will have a constant speed in a vacuum with respect to any reference-frame for spatial and temporal measurement?' It is with these questions that section 5.7 is concerned. But before attempting to answer these questions one must first understand what Einstein's explication of simultaneity amounts to. This will be the purpose of section 5.6.

5.6 EINSTEIN'S OPERATIONAL DEFINITION OF SIMULTANEITY

What is Einstein's explication of simultaneity?
Consider Fig. 20.

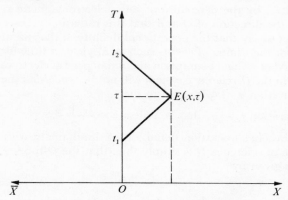

FIG. 20. *Einstein's simultaneity convention*

O is the origin of a spatial reference system, one dimension of which is represented by the line $\bar{X}OX$. The vertical direction on the page represents time. A temporal measurement system (i.e. a clock) at *O* measures out this time. So much for the time *at O*. But if we are to allocate times to events other than those that are spatially coincident with the clock at *O* (which we must do if our temporal measurement system is to be of any use at points other than *O* itself), then we must have a way of determining the *simultaneity* with respect to *O*, of events lying at some distance from *O*. For if our clock at *O* reads τ, say, then to say that some other event took place at time τ with respect to our frame of reference, is to say that this event is simultaneous in this frame of reference, with the event which is the clock at *O* reading τ.

To put it another way, event *E* takes place at *O*-time τ, if and only if event *E* is *O*-simultaneous with the clock at *O* reading τ. What is needed to make our time-measuring system complete is an operational definition (i.e. a way of determining), *O*-simultaneity.

Now let us assume that the speed of electromagnetic radiation (that is of *any* electromagnetic radiation whatsoever), is always the same with respect to *O*. Then equal *O*-distances will be covered in equal *O*-times by such radiation. Now any particular event *E* will be at some particular *O*-distance, say *x*, from *O*. Hence the *O*-time taken for some electromagnetic radiation to go from *O* to the position

x will equal the O-time taken for some electromagnetic radiation to go back again from x to O.

Let us assume that a ray of electromagnetic radiation leaves O at O-time t_1 (by the clock at O), and travels to the position x. Its arrival at x is the event E. Let us assume further that event E is also characterized by the departure of some electromagnetic radiation from x in the direction of O and that this radiation arrives at O at time t_2. Let us say that the operational definition, that we are going to adopt for the timing of events such as E, yields a time for E of τ. Then our Maxwellian assumption about the speed of electromagnetic radiation in the O-reference system will not be true *unless* the O-time duration $\tau - t_1$ is equal to the

O-time duration $t_2 - \tau$, that is, unless $\tau = \dfrac{t_1 + t_2}{2}$.

Now Einstein's operation definition of simultaneity with respect to a frame of reference O is simply this: that the O-time, τ, of any event E is given by

$$\tau = \frac{t_1 + t_2}{2}$$

where t_2 and t_1 are the arrival and departure times respectively at O of a ray of electromagnetic radiation making a round trip to the O-position, x, at which E occurs, the arrival and departure of the ray at x being coincident with the event E.

5.7 EXPERIMENTAL CONFIRMATION OF MAXWELL'S THEORY OF ELECTRO-MAGNETIC RADIATION

One conclusion that follows immediately from considerations of the last section is that Maxwell's theory of electromagnetic radiation is not true with respect to any spatio-temporal measurement system which does not determine simultaneity by a method, which yields the result that

$$\tau = \frac{t_1 + t_2}{2}$$

where τ is the time of an event, E; t_1 is the time that a ray of electromagnetic radiation would have to leave the system's clock to arrive at E when E occurred; and t_2 is the time that such a ray would arrive back at the clock after travelling from E.

But Einstein's operational definition defines τ as being equal to this quantity. Does it follow that there can be no experimental evidence that is contrary to the proposition that the speed of electromagnetic radiation is a constant with respect to any frame of reference, if we adopt Einstein's definition? If there can be such evidence, what form would it take?

The answer to the first question is 'Yes—there can be such experimental evidence.' The reason for this is that it is one thing to choose an operational definition for the purposes of some measurement system, and it is quite another for that operational definition to work. In section 3.6 it was stipulated that a distance-measuring system must be functional and what was meant by this was that given that the distance between *A* and *B* is purely spatial, then a distance-measuring system should yield some unique distance between *A* and *B*. Here the measurement in question is the measurement of the time of an event with respect to some particular frame of reference. Likewise, then, any system for such measurements will be satisfactory only if, for any particular reference-frame and for any particular event *E*, it yields the same time, τ, regardless of variations that one may make on the implementation of the system. Thus we should obtain the same result regardless of the wavelength of the radiation used, regardless of the speed with which the transmitter was travelling when the radiation left *O*, regardless of whether or not the radiation was reflected with a mirror at *E* that was moving with respect to *O*, and so on.

Now if Maxwell's laws of electromagnetic radiation are true with respect to some particular measurement system for spatial and temporal intervals, then Einstein's simultaneity convention will be functional within such measurement systems. So, conversely, if Einstein's simultaneity convention is not functional, Maxwell's theory of electromagnetic radiation is false for all physically possible measurement systems for spatial and temporal intervals. Failure of genuine attempts to demonstrate any lack of functionality in Einstein's simultaneity convention will therefore count as experimental evidence in favour of Maxwell's electromagnetic theory and hence in favour of the relativity of simultaneity, regardless of what measurement systems are adopted.

Gale in *The Language of Time*, p. 219, says 'It is difficult to see how a physical theory, such as relativity, which is concerned with spatio-temporal measurements, could inform us as to the manner in which the mind or consciousness is related to the physical world.' The above exposition has illustrated, I hope, how Einstein's theory, if confirmed, could inform us of at least one respect in which 'the mind is related to the physical world'. In our everyday experiences, simultaneity *seems* to us to be absolute. Einstein's theory could inform us that this appearance is an illusion. How this illusion (if illusion it is) comes about is in part a neuro-physiological and psychological question; whether or not it is an illusion is a question of physics.

I would guess that in so far as the authors Weyl, Eddington, and Grünbaum, mentioned by Gale in this regard, have said or implied that they believed that absolute simultaneity was mind-dependent, what they meant was simply that though simultaneity *seemed* to be absolute, they believed that it was not, in fact, absolute.

A myth, which is rather common throughout the literature on this subject, is that the notion of absolute simultaneity cannot be operationally defined with a definition which is functional in a world in which all signals are of finite velocity. Thus Gale (*The Language of Time*, p. 219) says 'The statement that two spatially separated events are absolutely simultaneous, i.e. simultaneous without any reference to a particular frame of reference, . . . lacks operational meaning in our world with its signals of finite velocity.' By 'lacks operational meaning', Gale presumably means 'cannot be operationally defined with a functional definition'. If this is what he means, however, what he says is very misleading. For whether or not all signals have a finite velocity, it still remains possible, all other things being equal, to define operationally absolute simultaneity. One could define it in terms of transported physically similar clocks that had been synchronized at the same place. Whether or not such a definition would be functional would depend only on whether or not similar clocks, that had been synchronized at one place and then transported to another place by different routes and perhaps at different speeds, were still in synchronism when they reached their destination. (It does indeed turn out that if such a system were functional, then Einstein's definition would not be and vice versa; though if Einstein's definition *is* functional, it can be shown that the functionality of simultaneity by transported clocks can be made as close to being functional as one likes, by reducing the speed of transportation.)

Another common myth is that Einstein's definition would be pointless or useless in a world in which indefinitely fast signals were possible. But nothing has been mentioned in this paragraph or in the subject-matter of other sections for that matter which eliminates the possibility or the impossibility of indefinitely fast signals. Contrary to popular mythology, it is quite consistent with Einstein's Special Theory of Relativity that there are things with finite real masses and energies travelling faster than the speed of electromagnetic radiation with respect to some frame or other. What is true is that the usual methods for determining simultaneity break down, once people have reason to believe that things can travel with the *same* finite speed with two different frames of reference which are in motion with respect to one another. It would matter to this not one iota whether or not there were *other* things that moved with respect to these frames at greater or even infinite speeds.

So much for the myths—what of the experimental evidence? To

traverse the intertwined skeins of both confused and rational thought about experimental confirmation of the Special Theory of Relativity over the last eighty years would require a very thick volume on its own. One reason for this is doubtless the conceptual innovation involved in the theory. Unfortunately the price of conceptual innovation is often confusion. But another, perhaps more important, reason for the complexity of progress in this area—all confusion aside—rests in disagreement as to what experimental evidence counts in favour or against particular theories. One reason for this is that it is perfectly possible for an experiment to be consistent with two inconsistent theories. For example, given that there are some ravens, the theory that all ravens are black is incompatible with the theory that all ravens under one year old are white. But both these theories are consistent with the experimental observation that this three-year-old raven is black. If there were any reason at all to believe that all ravens under one year old were white, then the observation of the three-year-old black raven could hardly be counted as evidence in favour of the theory that all ravens were black. It is clear then that what people will count as evidence in favour of the functionality of Einstein's simultaneity convention will depend on what other theories are at hand to explain the experimental results. Add to this the possibility of disagreement about the conditions which prevail throughout the experiment, and solid evidence becomes a very rare commodity. Furthermore this solidity will vary with history as we gain more and more knowledge about our surroundings and new theories are developed for consideration. An experiment which seems at the time to yield solid evidence in favour of a theory can, with hindsight, come to be thought irrelevant.

The two main theories which have competed over the years with Maxwell's theory of electromagnetic radiation and the functionality of Einstein's simultaneity convention have been:

(*a*) the aether theory of electromagnetic radiation,
and (*b*) Ritz's theory of electromagnetic radiation.

Both of these theories are based on Galilean kinematics. The aether theory is a wave-theory akin to the theory of sound propagation. Just as sound requires a substantial medium in which the propagation of the sound-waves occur, electromagnetic radiation, according to the aether theory, requires the presence of an 'aether' for its propagation. The 'aether' is supposed to pervade all of space, so that there are no vacuums in the sense that there is somewhere where there is nothing. The aether is always present if nothing else is.

Ritz's theory, propounded in 1911, allows that electromagnetic radiation can be propagated in an absolute vacuum, but with a velocity which is a constant with respect to the last object to transmit it.

One example will have to suffice to illustrate the remarks made above concerning experimental confirmation and disconfirmation. De Sitter's observations of double stars is widely held to refute Ritz's theory. De Sitter observed that a bright star in orbit about a dark star at first appeared on one side of the dark star, seemed to travel in front of the dark star, and then disappear around the other side. What else would one expect?

Well, if the speed of light was a constant with respect to its source, then we would expect the light leaving the bright star, as the star travelled towards us, to have a greater speed with respect to ourselves than it would have when the star was travelling away from us. Thus, depending on the speed of orbit of the light star about the dark star, and the distance of the double star from us, it should be possible for the 'fast' light originating on one side of the orbit to catch up with the 'slow' light of the previous orbit, thus presenting the observer with a double image of the bright star. But De Sitter failed to observe any such double images or corresponding distortions of the images of orbiting stars.

Do De Sitter's observations count as evidence against Ritz's theory? The assumption being made by those who think they do is that there is nothing between the double stars and the earth that would appreciably modify the speed of the light. But today it is believed that inter-stellar space contains much more material (charged particles and other matter) than people used to believe it to contain. Is the amount of this material sufficient, even on the basis of Ritz's theory, to bring the speed of the 'fast' light and the speed of the 'slow' light to the same value for most of the journey? If it is, then De Sitter's observations are consistent with Ritz's theory as well as Maxwell's.

The point of this section was to show how there can be experimental evidence that is contrary to the proposition that the speed of light is a constant with respect to any frame of reference: evidence which is applicable whether or not we adopt Einstein's simultaneity convention. That there can be such evidence indicates that the theory that simultaneity is absolute may be false and hence the theory of The Present may be false also. Many, if not most, physicists would claim, I would guess, that enough evidence is available for the reasonable rejection of a belief in absolute simultaneity. But the collection of such evidence will remain, so I predict, an activity which generates as much controversy as evidence, for some considerable time. Yet, even if there is no such thing as absolute simultaneity and Einstein's convention is applicable in this world, so that the theory of The Present is false, it may still be possible that some theory of relative past, present, and future is applicable. For any one observer, it may make sense to divide events into past, present, and future with respect to him. But it is hard to see what such a relativistic classification would possibly be based upon. Certainly the ontological differences of the theory of The Present would be lost. For any event that could be regarded as being in our future, there would be a reference-frame with respect to which that event is present and another in which it is past. Likewise Henry VIII's birth will be in the future for some reference-frame also. All events would take on an equivalent reality.

5.8 SOURCES AND NOTES FOR CHAPTER 5

Benjamin Whorf's article mentioned in section 5.1 is reprinted in *The Philosophy of Time*, edited by Richard Gale (Macmillan, London, 1968). An excellent bibliography of the debate on tense elimination entitled 'Works of interest mainly in connection with Part IV' may be found in Smart's reader *Problems of Space and Time* (Macmillan, New York, 1964), pages 443 and 444. This book, and Gale's reader mentioned above, contain many of the articles central to the debate. Gale carries on the debate in favour of tenses in his *The Language of Time* (Routledge & Kegan Paul, London, 1968) and Prior defends the retention of tenses and the importance of tense logic in his *Past, Present and Future* (Clarendon Press, Oxford, 1967), especially Chapter I. Smart's own position is to be found in Chapter VII of his *Philosophy and Scientific Realism* (Routledge & Kegan Paul, London, 1963). Quine's prime concern, as expressed in *Word and Object* (Wiley, New York, 1960), Chapter V, Section 36, seems to be that tenses are a logical inconvenience, a deviation from what he calls canonical form, but he allows himself the expressions 'now', 'then', 'before *t*', 'at *t*', and 'after *t*'. Reichenbach's prime concern was to investigate the logic of what he called 'token-reflexive' words such as 'I', 'now', 'here', 'there', and 'then' or what Quine, following Goodman (*The Structure of Appearance*, Harvard University Press, Cambridge, Mass., 1951, pages 290 ff.), called 'indicator words'. Reichenbach's treatment of such terms is to be found in his *Elements of Symbolic Logic* (Macmillan, New York, 1947), Section 50. This whole chapter and especially section 5.2 owes much, both directly and via other authors, to the writings of Smart, Quine, and Reichenbach.

The 'Theory of The Present' introduced in section 5.3 is a development of McTaggart's '*A*-series' theory with a view to a rendition that was at least, unlike McTaggart's, consistent. McTaggart's article, which is regarded by Gale and others as the origin of the tense elimination debate, is to be found in his *The Nature of Existence*

(Cambridge University Press, Cambridge, 1927), Book V, Chapter 33.

Moore's problem, with which section 5.4 is concerned, is raised again by Prior in his *Past, Present and Future*, mentioned above, p. 18. Moore's problem appears in *The Commonplace Book of G. E. Moore*, ed. C. Lewy (Allen & Unwin, London, 1962), Notebook II (C 1926), entry 8, p. 97. Moore's article about the logic of 'exist', from which the examples about the tame tigers existing are derived, is to be found in the article 'Is Existence a Predicate?', published in *Logic and Language*, second series, ed. A. G. N. Flew (Basil Blackwell, Oxford, 1953).

Sections 5.5 and 5.6 are largely derived from Part I of Einstein's *Relativity: The Special and the General Theory* (15th edn., Methuen, London, 1954). The discussion of the possibility of experimental confirmation of Maxwell's theory of electromagnetic radiation in section 5.7 was stimulated by many sources. Three of the main stimulants have been Richtmeyer and Kennard's *Introduction to Modern Physics*, Chapter IV, 'The Theory of Relativity' (McGraw-Hill, New York, 1947), Adolf Grünbaum's 'Logical and Philosophical Foundations of the Special Theory of Relativity', in *Philosophy of Science*, edited by Danto and Morgenbesser (Meridian, New York, 1960), and Einstein's *Relativity: The Special and the General Theory*, mentioned above.

Einstein on page 17 of that book seems to accept De Sitter's observations unquestionably as disconfirming evidence for Ritz's theory and as confirming evidence for Maxwell's theory, and many authors seem to have followed his example. But in the preface Einstein warns that he has 'purposely treated the empirical foundations of the theory in a "step-motherly" fashion, so that readers unfamiliar with physics may not feel like the wanderer who was unable to see the forest for trees'. While on this topic, I must in passing criticize Grünbaum's uncritical acceptance (page 430 of Danto and Morgenbesser) of the significance of R. Tomaschek's experiment (*Annalen der Physik*, 4th ser., vol. 73 (1924), p. 105) in disconfirming Ritz's theory. Tomaschek used a Michelson–Morley apparatus which was supposed to use stellar radiation as a source. The so-called stellar radiation, however, had passed through the earth's atmosphere, a window, an optical collimator and had either been reflected or transmitted by a half-silvered mirror before entering the apparatus. Ritz's theory is *not* that electromagnetic radiation has a speed which is a constant with respect to its source *regardless of the media through which the light subsequently travels*.

CHAPTER 6

Temporal Asymmetry

6.1 INTRODUCTION

The idea of time-flow may be regarded as a combination of:

(a) the theory of ontological disparity between present events as opposed to past or future events—a theory that I have dubbed 'the theory of The Present'; and

(b) a theory that time is fundamentally asymmetrical in some sense that guarantees a profound difference between the past and the future.

Chapter 5 was concerned primarily with (a). This chapter is concerned with (b). Verbal labels and specially pleading terminologies (such as tenses) aside, there was nothing in the last chapter to suggest that the difference between past and future is any more fundamentally asymmetrical than the difference between north and south. One may wonder, however, if there is any point in looking for the asymmetry between past and future if events cannot be categorically classed as being past or future, as modern trends in physics seem to be indicating. But the future was supposed to come later than the past and the past to come earlier than the future and it is perfectly consistent with the acceptance of Einstein's theory that there is some fundamental difference between event A being earlier than event B for some reference-frame, and event A being later than event B within that same reference-frame. So the question arises: with respect to any reference-frame for the assignation of distances and times, what relational quality, if any, does some event's being earlier than some other event invariably have that some event's being later than some other event invariably lacks? In answering this question one must beware of all the pitfalls that were pointed out in section 4.2, when the question was raised as to what qualitative differences there are between time and space. In particular one must be on the lookout for special pleaders. We know that there are two directions of time. What one wants to know is what the qualitative differences between them are, if indeed there are any. The next few sections will consider some attempts to answer this question.

6.2 CAUSES AND EFFECTS

One attempt to answer the question at the end of the preceding section lies in pointing out that causes always precede or are simultaneous with their effects but that effects never precede their causes. When one points out, however, that one would not call some event or object, *B*, a cause of some other event or object, *C*, if *C* were thought to be wholly earlier than *B*, it appears that the notions of 'cause' and 'effect' are special pleaders with respect to temporal asymmetry. In this section I wish to bring attention to the extremely widespread effects that this has throughout our language. I shall then argue that even if we strip the notions of 'cause' and 'effect' of their directional bias, there is no evidence to suggest that the resulting causal relation is always exemplified asymmetrically in time.

Our language is enormously rich in concepts which are logically related to the notion of causation. The verb 'to do' yields an example. Anything that does something is something which *causes* what it did to occur. The verb 'to do' is, in turn, logically related to very many other words in our language—so many, in fact, that it is a common fallacy among primary-school teachers (at least it used to be so) to describe any verb whatsoever as a 'doing' word. There are obvious exceptions to this rule, of course, not least of which is the verb 'to be'. If something *is* a stone, it does not follow that it is doing anything. Nevertheless the great majority of sentences uttered in normal everyday conversation, or uttered by a television announcer, or occurring in a novel or an average magazine have their causal connotations. Considerations of tenses aside, these causal connotations which pervade our everyday discourse bear with them the implication of something being earlier than something else or, if not that, the possibility of something being earlier than something else. But it is just because they bear these implications that they are of no use in yielding some significant quality invariably obtaining in the instantiation of an 'earlier than' relation that does not obtain in any instantiation of a 'later than' relation. Thus no such qualitative difference is indicated by the truth: 'If *B* is earlier than *C*, then *B* is not the action of which *C* was the corresponding intention', if it is known that we would not, without misusing the language, call *C* an intention to perform the act *B*, if we know that *B* was earlier than *C*.

Some philosophers have seemed to believe that the fact that our language is crammed with these logical asymmetries is some sort of indication of some significant qualitative difference between earlier and later. Thus Gale in *The Language of Time*, page 104, writes:

Any one of these logical asymmetries, when viewed in isolation from the others, will seem to be a rather unexciting tautology. . . . However, when a synoptic view is taken of them they no longer will appear trivial or unimportant. It will then be seen that these logical asymmetries interlock and support each other like the legs of a tripod, such that one of them cannot be jettisoned without giving up the others as well.

But there is no question here of jettisoning any of the special pleaders. It is just a matter of recognizing them for what they are. No matter how many of them there are and no matter how much they logically interlock and 'support each other', whatever that may amount to, they still remain special pleaders and cannot, therefore, be used to display to us any qualitative rather than mere numerical difference between the two directions of time.

There has been in the literature a considerable debate in recent years about whether or not it is logically possible for an effect to precede its cause, and many a philosopher reading this will no doubt object to my bland assertion that it is not. But this whole verbal debate is best short-circuited by allowing that two events, B and C, may be related in a way that *would be* a case of B causing C *except that B is later than C*. So let us allow Reichenbach, Dummett, and others a sense of 'cause' and 'effect' such that in *this* sense, it is logically possible for an effect to be earlier than its cause, except that for this sense of 'cause' and 'effect', *we* shall use the words 'sub-cause' and 'sub-effect'; and we shall *retain* the words 'cause' and 'effect' to be used in such a way that it is logically necessary that:

> B is a cause of the effect C if and only if
>   (a) B is a sub-cause of the sub-effect C
> and (b) B is earlier than C.

The concept of 'is a sub-cause of' cannot now be a special pleader with respect to the earlier to later direction of time. If someone now claims that the direction of time from earlier to later invariably has the quality of being the same direction as the direction from B to C where B is any sub-cause of its sub-effect C, then this will at least be informative. But will it be true? That will depend on what one considers to be left of the notion of causation, once it has been stripped of its temporally directional connotations.

What would be left for most philosophers, I guess, would be the notion of natural or physical necessity—so the next job is to explain this notion. Once again, volumes would be needed to examine this notion in any detail, but it will suffice for the purposes of this section if we say briefly that:

An event E is physically necessary for another event F if and only if the occurrence of F without the occurrence of E, with everything else remaining the same, is contrary to a law of nature.

If E and F were both in the past, then, one might express this by saying that all other things remaining the same, if E had not occurred, F would not have occurred. This makes it look as though E was the cause of F. For example, if the butler's hatred of Sir Wilfred was the cause of his killing Sir Wilfred, we might express it thus:

'If the butler had not hated Sir Wilfred, then the butler would not have killed Sir Wilfred.' Or:

'Since the butler killed Sir Wilfred, the butler must have hated Sir Wilfred.'

Is it the case then, that causes are always physically necessary for their effects? No. Sometimes it is the other way around. Suppose the butler was rather over-enthusiastic. He not only thrust a knife into poor Sir Wilfred's heart, but he also hit him on the head with a blunt instrument and smothered Sir Wilfred's face with a pillow. The coroner later finds that the cause of the death of Sir Wilfred was the knife entering the heart. What is meant by this is not that Sir Wilfred would not have died if the knife had not entered the heart, for that would almost certainly have been false. Rather what is meant is that if Sir Wilfred were alive and well, there would have been no occurrence which was a knife's entering his heart. That is, consistent with the definition of physical necessity, the effect in this case, namely Sir Wilfred's death, is physically necessary for the cause. If this sounds a little bizarre, it is probably because we usually say in such cases that the cause was *sufficient* for the effect, but I do not think there is much to be gleaned from the fact that our language is biased in this way. More strictly speaking we should say that the cause was *physically sufficient* for the effect; where physically sufficient is defined thus:

E is physically sufficient for F if and only if F is physically necessary for E.

Thus causes can be either physically necessary or physically sufficient for their effects, and the same would apply, of course, to sub-causes and sub-effects, since the notions of physical necessity and physical sufficiency as they have been defined here contain no connotations of something's being earlier than or later than any other thing. But if there is no temporal bias in the notions of sub-cause and sub-effect, it is hard to see what semantic difference there remains between the two expressions. Sub-causes can be physically necessary for their sub-effects, but so can sub-effects be physically necessary for their sub-causes. Indeed, unless sub-causes can be differentiated from sub-effects by sheer inspection, there seems to be no basis for differentiation here at all. Let us assume that we had witnessed Sir

Wilfred's killing at the hands of the butler. We might claim to have witnessed the cause of Sir Wilfred's death. Philosophical difficulties about observing causes on one side, what we have observed, if anything, is the sub-cause of Sir Wilfred's death and that this sub-cause was earlier than the sub-effect. But is this any different from asserting that we observed that the butler's attack and Sir Wilfred's death were causally related (one way or another) and that Sir Wilfred's death came *later than* the butler's attack? What difference, *besides* the temporal difference of being later than the butler's attack, does Sir Wilfred's death display in its causal relationship to the butler's attack? If being a sub-cause is to be either a physically necessary or a physically sufficient condition for a sub-effect, and being a sub-effect is to be either a physically necessary or physically sufficient condition for a sub-cause, it would appear that we might just as well have dubbed Sir Wilfred's death as a sub-cause of the sub-effect which was the Butler's attack.

If this is so, then it would simply be false to say that all sub-causes precede their sub-effects. For any case of a sub-cause preceding a sub-effect could equally be described as a sub-effect preceding a sub-cause.

I conclude therefore that the statement with which this section began is correct. We know nothing substantial about the difference between earlier and later if all we know is that causes are earlier than their effects and effects are later than their causes.

6.3 KNOWLEDGE AND DECISION

Much is often made of the fact that our knowledge of the future is very different from our knowledge of the past. It is relatively easy to find out in particular cases, given present evidence, who was born when, who married whom, and who died, and under what particular circumstances these events took place. It is not so easy to know who is going to be born, who is going to marry whom, who will die and when, and under what circumstances these events are going to take place. There seems indeed to be much *more* knowledge of the past than there is of the future. We can have knowledge of the past by remembering what happened or by examining historical records. But we do not memorize future events, there are no traces of future events that precede them, and there are no records of the future.

In this section I claim that there is an important difference between knowledge of the past and knowledge of the future, but I shall further argue that this difference is in no way indicative of a qualitative difference between the direction of time from earlier to later and the direction of time from later to earlier. Indeed it will be argued that the concept of knowledge is itself a special pleader.

First, however, the meaning of 'knowledge' must be examined.

It was traditional in philosophic circles in the thirties, forties, and fifties to analyse '*X* knows *p*' where *X* is some person and *p* is some proposition as follows:

X knows *p* if and only if
(*a*) *p* is true;
(*b*) *X* believes *p*;
(*c*) *X* is justified in believing *p*
   (or *X* has good reasons to believe *p*)

Gettier, in his article 'Is Justified True Belief Knowledge?', in *Analysis*, vol. 23, no. 6 (June 1963), pointed out that these analyses are unsatisfactory on the following assumptions.

  (i) If some proposition *q* logically implies another proposition *p*, then if one is justified in believing (or has good reasons to believe) *q*, one is justified in believing (or has good reasons to believe) *p*.
 (ii) It is perfectly possible to be justified in believing (or to have good reasons to believe) a falsehood.

The first of these assumptions is uncontroversial; the second is rendered plausible by a single example. An immigrations officer may be perfectly justified in believing (or have good reasons to believe) that a passport is genuine, even though the passport is in fact forged —provided, of course, that the forgery is a very good one.

Now add to (i) and (ii) the fact

 (iii) that it is possible validly to deduce a truth from a falsehood.

As an example of this consider:

   I am a female
∴   I am not a bachelor.

The argument is valid, the premiss is false, and the conclusion is true.

Now given these three assumptions, it follows that someone *X* could be justified in believing (or have good reasons to believe) some falsehood *q*, from *q* validly deduce some truth *p*, and thereby believe and be justified in believing (or have good reasons to believe) *p*.

Thus *p* would be true,
     X would believe *p*,
and   *X* would be justified in believing (or have good reasons to believe) *p*.

For example, let the immigration officer be X. X examines Jones's passport and comes to believe, justifiably, that this is a genuine passport of Jones. Little does he know that it is nothing of the sort. Nor does he know, that as well as being in possession of this document, Jones is also in possession of a genuine passport that he is carrying hidden in his briefcase.

But from 'This is a genuine passport of Jones' (q), X deduces validly that 'Jones is the possessor of a genuine passport' (p).

On the analysis of 'X knows that p' given above, we would have to agree, in this case, that X knew that Jones was the possessor of a genuine passport. But X did not know that Jones was in possession of such a passport. The usual analysis of knowledge therefore fails.

It is of interest here to ask oneself why X's true belief that Jones was the possessor of a genuine passport was not a case of knowledge. The reason seems clear enough. X's belief came about as a result of a deduction from a false premiss, and deductions from false premisses are *not likely* to result in true beliefs. That is, the rejection of X's belief as a case of knowledge is that the *cause* of the belief was such that, under the circumstances, the belief was not likely to have been true. A better analysis of 'X knows that p' would therefore seem to be:

X knows that p if and only if

   (i) p is true

   (ii) X believes p

and (iii) X's belief was caused in such a way that it was likely to be true.

or to put the third clause in a way that might appeal to a modern communications engineer:

   (iii) X's belief was the end-product of a relatively noise-free causal process.

This analysis is in many ways unsatisfactory to the analytic philosopher, for the probabilistic notions of 'noise-free causal process' or 'being caused in such a way that it was likely to be true' are, if anything, more semantically mysterious than the notion of knowledge itself. Indeed, it is not at all clear to me that the probabilistic notions involved here are not cognates of 'knowledge' anyway. But all that is by the way. An analysis may be unsatisfactory in this way, yet nevertheless be both correct and insightful. And the insight I wish to bring into the open here is that the third clause of the above analysis demands that the belief involved in a particular case of knowledge is the end-product of a certain sort of causal process. The belief has to be an appropriate sort of *effect*. But it was argued in section 6.2 that 'cause' and 'effect' were logically biased with respect to 'earlier' and 'later'. If this is so, it would appear that 'knowledge' too is logically biased in this way. I shall now show how this logical bias, this special pleading for temporal asymmetry inherent in the concept of knowledge, explains why knowledge of the future is different from knowledge of the past and why it is so much easier to come to know the past than to come to know the future. These differences, it will be shown, yield no qualitative differences whatever between the two directions of time.

Consider, for example, the case of knowledge or some past event, E, which is brought about via a memory of E, or by examination of traces or records of E. In all such cases there is a causal chain leading from the event whose occurrence is known to the belief in the occurrence. (See Fig. 21.)

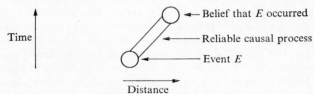

Fig. 21. *A belief reliably caused by a memory trace,*
or other trace or record of a past event

The temporal inverse of this situation is not necessarily a case of knowledge at all; it is a belief which reliably causes an event E, the belief being that the event E will occur. (See Fig. 22.)

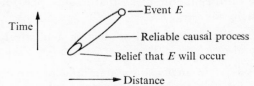

Fig. 22. *A belief that reliably causes the event*
which is believed to be going to occur

Such situations are commonplace. For example, if a person aims at achieving some end, his belief that the end will be achieved is causally essential, more often than not, to his subsequent behaviour, which in turn brings about the event aimed for. It is not that all action depends on a belief in the success of that action. Sometimes we act in hope rather than in certainty, especially when the stakes are high. But with respect to our everyday behaviour, our beliefs that our actions will bring about the desired effects are usually causally important in the event of our actions actually bringing about the effects. In such cases, it is true, we would more often than not say that the agent knew that he would produce the event which was his aim, not because of the reliable process that linked his belief to the end result, but rather as a result of remembered past experiences of like situations, which caused in him this self-confidence. The point here is that such experiences could be lacking—in which case it would be correct to say that the agent's belief was not a case of knowledge, and yet the agent's belief nevertheless reliably caused the believed in event to occur.

Also, beliefs in the occurrences of future events can reliably cause those events in cases where there is no intention to bring about the event, and these beliefs, too, may or may not be cases of knowledge, depending on their temporal antecedents. For example, there is the well-known 'Oedipus effect' as Popper has called it, an example of which is as follows:

A leading and well-respected member of the Stock Exchange believes that the stock market is about to decline. Being a loquacious gentleman, he cannot of course keep this to himself. When the 'news' leaks out, people rush to sell their stocks. Supply exceeds demand and the prices of the stocks fall. Let us assume that the cause of the stockbroker's belief was a misheard conversation in a crowded bar-room between two public servants discussing the bank-rate. Knowing this to be the cause of his belief and knowing that he misunderstood the public servants' conversation, we would not classify his belief as knowledge.

So the temporal inverses of cases of knowledge of the past are not necessarily cases of knowledge of the future. A belief about the future, reliably causing the believed-in event to occur, does not in itself guarantee a state of knowledge, whereas a belief about the past being reliably caused by the event which has occurred is invariably regarded as being a state of knowledge.

I conclude that 'knowledge' is a special pleader with respect to temporal asymmetry, and that no qualitative difference between 'earlier' and 'later' can be shown to follow from differences between our knowledge of the past and our knowledge of the future. The strict temporal inverses of knowledge of the past are simply not counted as necessarily being cases of knowledge of the future. Even if the temporal inverse of retrodiction could be regarded as prediction, the temporal inverse of a reliably caused retrodiction is *not* a reliably caused prediction. Rather it is a prediction which itself reliably causes what is predicted. If it could be shown without special pleading that there are important asymmetries between knowledge of the past and the strict temporal inverse of knowledge of the past, on the other hand, there would be reason to believe that there was a qualitative difference between the two directions of time. But I know of no attempt to show that such differences obtain.

6.4 DOES THE SECOND LAW OF THERMO-DYNAMICS DISPLAY A TEMPORAL ASYMMETRY?

In section 4.4, it was shown that there were non-numerical differences between space and time exhibited by various laws of nature. In these laws, it was not possible to interchange time with one of the spatial dimensions. Can one in a similar way locate a non-numerical difference between the two directions of time? Is there some law which refers directly or indirectly to a time-change Δt in the direction from earlier to later, such that if $-\Delta t$, representing a time change from later to earlier, was substituted for Δt throughout the law, the law would no longer hold?

Most laws of physics are not of this nature. Thus Newton's laws of dynamics given the appropriate initial conditions determine the motions of the planets as we observe them to all intents and purpose. Yet if all these motions were reversed, the resulting motions would still be in accordance with Newton's laws. Indeed over the years, the only theory enjoying much support, which has been thought to exhibit temporal asymmetry, is the Second Law of Thermodynamics. The law may be stated in a number of ways, which, given other assumptions usually made within the theory of thermodynamics, turn out to be equivalent. One such way is as follows:

Heat cannot of itself, i.e. without the performance of work by some external agency, pass from a cold to a warmer body.

That is, given two bodies causally isolated from everything except one another, one of which is at a higher temperature than the other, any flow of heat from one to the other will be from the warmer to the cooler and so will be such as to tend to equalize the temperatures of the two bodies.

An alternative and rather more general way of stating the law is in terms of a quantity called entropy which must now be defined. Assume that a body is at some temperature T, and that it receives some quantity of heat ΔQ which is small enough not to cause an appreciable change in the body's temperature. Then the change in entropy of the body (written ΔS) is equal by definition to $\Delta Q/T$. The Second Law of Thermodynamics may now be expressed as follows:

Given any causally isolated system of bodies between which there is an interchange of heat, then the total change in entropy is always positive.

To illustrate this consider an isolated system consisting of two bodies B and C. B is at a temperature T_B and C is at a temperature T_C, such that T_B is greater than T_C. According to the first way of enunciating the Second Law, heat will, if anything, flow from B to C. Let a small quantity ΔQ flow from B to C, i.e. *out* of B and *in* to C. B's entropy will drop by a quantity $\Delta Q / T_B$, and C's entropy will rise by the quantity $\Delta Q / T_C$. But since T_B is greater than T_C, the rise in C's entropy will be greater than the fall in B's entropy. The total entropy of the system will therefore increase.

Obviously the law, on the surface of it, exhibits a temporal asymmetry. Given a particular situation, heat can flow one way, but not the other. It can only flow in a direction which results in an over-all increase in entropy.

Two questions arise:

   (i) Is the 'law' universally true; that is, is it really a law?
   (ii) If it is a law, does it really, on closer analysis, display an asymmetry in time?

Let us look now at the first of these questions. Certainly common experience seems to support the law extremely well. No one has ever experienced the phenomenon of sitting in a lukewarm bath, and suddenly finding to his surprise that one end of the bath grew hotter at the expense of the other end, with the result that the water about his toes started to boil while the water behind his back started to freeze. Yet it is a common enough experience that the hot water and cold water poured into a bath interchange heat in such a way to produce a more homogeneous distribution of temperature.

But let us consider such an event at a sub-microscopic rather than a macroscopic level. The temperature of a body is these days thought to be merely the mean kinetic energy of its atoms or molecules. If B is at a higher temperature than C, then the mean kinetic energy of B's molecules is greater than the mean kinetic energy of C's molecules. When the bodies are brought into contact, B's molecules, via collision or other mechanical interactions, give up some of their kinetic energy to C's molecules, and thus the heat flows from B to C. But all these sorts of sub-microscopic interactions are thought to be reversible. That is, if it were possible to take a moving picture of any one of these interactions and play the picture backwards, the displayed phenomenon would be perfectly consistent with other laws. But if this is the case for any one interaction, it should be the case for any number of them or, for that matter, all of them. But if the aggregate of all the interactions is a reversible process, then it would appear that heat *can* flow from a cold body to a hotter one unaided by any external agent. Thus the Second Law of Thermodynamics is inconsistent with the time-reversible theories governing the inter-actions of the sub-microscopic constituents of matter.

Boltzmann, in 1877, suggested a reinterpretation of thermo-dynamics and the Second 'Law' of Thermodynamics in particular, to accommodate both the time-reversibility of the laws of sub-microscopic interactions and the everyday macro-effects which we observe when, for example, we take a bath. The idea is this. Given any way of subdividing the possible states of an individual molecule with respect to position and velocity, then given the vast number of molecules in the bath, the probability of the macroscopically more homogeneous states of the bath with respect to temperature is greater than that of the less homogeneous states. A state of the bath with all the hot molecules 'stacked' at one end and all the cold molecules 'stacked' at the other end would be most highly improbable compared to one wherein the molecules were more equally dis-tributed. The situation is roughly like the comparison between a pack of cards with all the aces on top, followed by the kings, the queens, etc., down to the twos in that order, with a pack in which the order is more 'random'. If we started off with the stacked pack and pro-ceeded to shuffle it, then the odds in favour of the pack ending up in a random state would be very great indeed. Likewise if we started off with a random distribution of the cards in the pack, and then shuffled the pack, the chances *against* ending up with a stacked pack are once again very great. Correspondingly, if we regard the sub-microscopic interaction of the molecules as a shuffling process (in which case we must regard the temporal inverse of these interactions as a shuffling process also), it is clear that if we start off with a collection of stacked molecules (that is with hot water at one end of the bath and cold water at the other end) and allow the 'shuffling' to proceed, we shall very likely end up with a random distribution of molecules, that is with lukewarm water everywhere. Again, were we to start off with the random distribution, that is with the lukewarm water everywhere, and allowed the sub-microscopic shuffling to proceed, the chances *against* the water coming to boil at one end of the bath and coming to freeze at the other end, would be very great. Thus, rather than saying now that in a causally closed system heat *always* flows from the hotter of two bodies to the colder of the two, one says instead that in a causally closed system it is very much more likely that heat will flow from the hotter body to the colder body rather than the reverse. Or, in terms of entropy, one could say that it is always more likely than not, given any change at all within causally isolated system, that total entropy will increase.

This reinterpretation of the Second Law of Thermodynamics, however, may still seem to be temporally asymmetric. Indeed many

physicists during this century have taken the probabilistic interpretation of thermodynamics as differentiating between the two
directions of time. But this is a mistake. Let us call a stacked or
ordered state of a system, '*O*', the process during which the system
is being shuffled, '*S*', and a random or disordered state of a system,
'*D*'. Then what has just been argued is that, given *O* followed by *S*,
the odds are great that the result will *then* be *D*. This then is compared
with: given *D* followed by *S*, the odds are *very low* that the result will
then be *O*. These two statements are true, but do *not* describe temporally inverse situations.

To make this clear, let us express the two statements as follows:

(i) The odds that *D* will occur, given that *O* is now the case and
that *S* is now about to begin, are great.
(ii) The odds that *O* will occur, given that *D* is now the case and
that *S* is now about to begin, are small.

Now the true temporal inverses of (i) and (ii) are generated by
reversing the tenses of the two statements. This yields:

(iii) The odds that *D* has occurred, given that *O* is now the case
and that *S* has been going on up to now, are great.
(iv) The odds that *O* has occurred, given that *D* is now the case
and that *S* has been going on up to now, are small.

In other words, if nothing else is known except that a pack of
cards has been genuinely shuffled and the result is a stacked pack,
then the chances are that the cards were initially disordered. Again,
if as a result of a genuine shuffle, a disordered pack results, then the
chances are remote that the pack was initially stacked.

In the case of our bath of water, *given* that the water has been
causally isolated and that all that has been going on has been
molecular shuffling, then the odds would be great, *given* that the
bath water is hotter at one end than the other, that it was previously
in a homogenous state. And given that it is now homogenously
lukewarm, the odds are small that the water at one end was previously noticeably hotter than the water at the other end. In so far
as these statements are counter-intuitive it is because we tend to
confuse two sorts of probability—statistical probability and epistemic
probability. Statistical probability (which is the sort of probability
being used by Boltzmann) is a function of the number of possible
states of a system under various descriptions of the system. Thus
when we say that the probability of two dice showing seven is one in
six, we are saying that the ratio of the number of possible states
wherein the dice are showing seven to the total number of possible
states is as one is to six. Epistemic probability, on the other hand, is

very much a function of what is known about the system, the origin and history of systems that are like it, and so on. Thus the *statistical* probability of Cedarlegs winning the Randwick handicap in a field of twelve horses is one in twelve. But for a good punter, having studied his form guide over the previous year or so, the epistemic probability may be one in three. For the trainer, who has instructed the jockey to give the horse a bit of a rest, the epistemic probability of Cedarlegs winning is close to zero.

Now if knowledge is *necessarily* temporally asymmetrical as I have argued in the previous section, then it is clear that epistemic probabilities about future events given present and past events, will be very different from the corresponding epistemic probabilities about past events given present and future events. So in considering our examples involving shuffled packs of cards, and bathtubs filled with interacting water molecules, we must eradicate from consideration any knowledge we possess about how packs of cards come to be in stacked states as a rule or how baths of water usually come to have hot water at one end and cold water at the other. The easiest way to do this is to imagine that the cards have been being shuffled for an eternity and that the water has been causally (and hence thermally) insulated from everything else for an eternity. During such immense periods of time all sorts of unlikely sequences of the card pack will arise at infrequent intervals, but by the same token, if Boltzmann's account is correct, many, though infrequent, cases of inhomogeneity in temperature will come about in the bathtub. It is in consideration of such infrequent events such as these, together with their likely and unlikely preceding and succeeding counterparts that statements (i), (ii), (iii), and (iv) are seen to be equally plausible. But these statements are more than merely plausible. For if what we mean by a system being in a disordered state is that the system is in a state that is classified as one of a majority of possibilities, then all of (i), (ii), (iii), and (iv) become analytically true, and hence can tell us nothing of the way the world happens to be, let alone anything about how one direction of time differs qualitatively from the other.

6.5 LOCAL AND ACCIDENTAL ASYMMETRIES WITHIN TIME

If we came across a lukewarm bath of water, we would assume, contrary to the probabilities yielded by statistical thermodynamics, that it had recently arrived in that state as a result of some cold and some hot water mixing together. This is because we know that most of the water available in modern houses is either cold or hot, depending on which tap is turned. Also we know that people are not in the habit of keeping water in bathtubs for very long periods, let alone keeping their bathtub in thermal isolation. In short, we are continually observing situations where low entropy states of a small system are being set up by some sort of intervention with the system, arising from a greater external system. The hot water itself is a product of such an intervention. It results from an input of heat energy from an electrical heating element into water which was originally cold. The heat of the heating element arises out of another intervention, the passing of electrical current through the element. The electrical current is generated in a power-house which is a mechanism which takes coal and burns it in air to produce steam to drive turbines. The coal itself was originally photosynthesized by plants using sunlight and the carbon dioxide of the atmosphere. The solar system itself was derived from the collapse of a cloud of dust carried in one of the spiral arms of our galaxy. Perhaps the galaxy itself is a branch-system of some even greater system. Now most of the thermodynamic branch-systems that we observe with rising entropic states are a direct result of the flow of energy outward from the sun or outward from the hot core of the earth into a relatively empty space. Every human body and the branch-systems within it are part of this process also, and it has been conjectured by Grün-baum and others that our consciousness of a 'flow' of time is caused by the entropy increases within these branch-systems of ours. The details of how this consciousness comes about are not only beyond the scope of this book, but are also beyond the scope of man's knowledge at this time. Suffice it to say that this area of knowledge (or the lack of it) carries with it the maze of philosophic problems associated with theories of perception and the relations between bodily processes and mental events. Yet it is because our consciousness of what seems to be a temporal asymmetry is as likely as not to be caused by a local pattern of energy flux, that it is important not to assume that this apparent asymmetry is real, in the way in which one might wish to treat the existence of matter as real—a fact which is immediately given by the senses. A sleeping shark must lie on the bottom with its gills facing into the current if it is not going to suffocate. For all we know, it experiences the upstream orientation very differently from the downstream orientation, and in a sense it would experience both a spatial and temporal asymmetry. After all, the water is *in front of* its gills *earlier than* when it is *behind* its gills. But this is an asymmetry of the flow of water *within* space and time—not an asymmetry *of* either space or time themselves. If we said otherwise, we would have to allow that the movement of a perfect sphere from left to right indicated an asymmetry of both space and time—a spatial asym-metry because it was to the left earlier than it was to the right, and a temporal asymmetry because earlier it was more to the left than it was later. Similarly, it would seem to be improper to say that time was asymmetrical on the basis of observations made within a stream of energy which we know to be causally operative in generating the observed temporally asymmetric situations.

However, cases of water flowing in all sorts of directions are commonplace and cases of spheres moving from right to left are just as probable as cases of spheres moving from left to right. But it seems highly improbable that radiation should focus in from the depths of space on to a galaxy resulting in a system more or less like ours except that the common processes were all reversed. Thus the incoming radiation would have to activate the air molecules in the vicinity of a hot log in such a way as to send particles of carbon and molecules of carbon dioxide and water vapour that happened to be there down towards the log just in time to receive radiation in the visual wavelengths, which would heat them to a temperature just sufficient to cause their chemical reduction into cellulose just as they met the log. Furthermore, these extreme coincidences would continue until the log was completely re-constituted whole and the chemical absorption of the incoming radiation would be so efficient as to leave the log quite cold. Of course a set of coincidences like this would not be isolated. The whole galaxy would be filled with them. If, in such a galaxy, there were sentient beings rather like ourselves on some planet or other, their internal processes would also be reversed, including all those that constituted their reasoning and experience. They would therefore be reverse-thinking, as like as not, that their galaxy seemed highly asymmetric in time and would be reverse-thinking that it would be highly improbable for a galaxy to operate in a temporal direction that was opposite to their own (such as ours does). Would they be wrong? Would they not be able to see that the majority of galaxies worked in the same direction as ours and that they were the odd men out? No, they would not. For, with respect to *their* direction of time, our galaxy would be absorbing the radiation it emits with respect to our direction and would only be emitting the relatively weak radiation it absorbs with respect to our direction. Our galaxy, to them, would appear only as part of the general darkness of the night sky. Their galaxy would appear to us in a similar fashion.

However, if there were two sets of galaxies in the universe, one operating in one direction and the other operating in the reverse direction, then with respect to either one of these temporal directions one set would be transmitting radiation outwards and one set would be the absorbing recipients of radiation travelling in towards them. But if this is the true picture, the following question arises: from whence comes the radiation that is centred in on the absorbing galaxies? How could it be that radiation travelling in from in-definitely great distances would be focusing itself on these galactic pin-points in space?

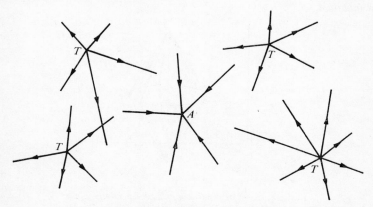

Fig. 23. *An absorbing galaxy surrounded by a group of transmitting galaxies*

Figure 23 illustrates the difficulty. It shows an absorbing galaxy (*A*) surrounded by a group of transmitters (*T*). At any point in space, the radiation coming from a transmitter will be *diverging*, but the radiation going to an absorber will be *converging*. Without further assumptions, it would appear that the radiation being absorbed could not be the radiation being transmitted.

But why should the idea of radiation travelling in from *indefinite* distances and focusing itself on a reverse-operating galaxy be so objectionable? Grünbaum, on page 269 of his *Philosophical Problems of Space and Time*, seems to think that one reason for this is that we never encounter such processes. But this will never do for, as pointed out above, were we creatures living in a galaxy in which this did occur, we would experience the incoming radiation as radiation which our galaxy was transmitting outwards. Another reason given and on which Grünbaum, Hill, and Popper are all agreed (see Grünbaum's *Philosophical Problems of Space and Time*, Chapter 8B), is that, as Popper remarks (*Nature* (1957), p. 1297), 'Only such con-ditions can be causally realised as can be organised from one centre ... causes which are not centrally correlated are causally unrelated, and can co-operate only by accident. ... The probability of such an accident will be zero.'

Popper, then, is not averse to the serious consideration of the sort of occurrences illustrated in Fig. 24 where an event *A* causes an infinite set of events *C*, distributed widely in space, which in turn result in converging processes which yield the effect *B*. Nor does he object to the sort of processes illustrated in Fig. 25 wherein a cause *A* generates an infinite set of diverging processes that continue *ad infinitum*.

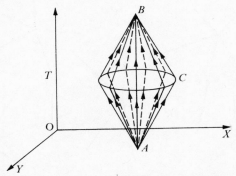

Fig. 24. *A set of conditions,* C, *which jointly cause an event,* B, *having themselves been 'centrally correlated' by the cause* A

What he does object to is the temporal inverse of Fig. 25, namely the occurrences illustrated in Fig. 26, wherein an infinite set of processes with no common causal origin converge to cause the event *A*. The claim is that this cannot occur—its probability is zero.

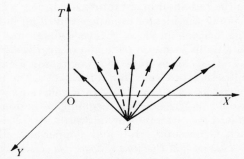

Fig. 25. *Event* A *causes a diverging and never-ending set of processes*

But this will not do either. The fact is that at any point in space, the conditions at any particular time at that point (for example the electric and magnetic field strengths) are the result of radiation that has converged on that point from every spatial direction. So such occurrences are not rare, let alone nonexistent. On the contrary they are ubiquitous. What is far rarer, but not all that rare, is the situation shown in Fig. 24. Such occurrences could include, for example, an image being formed by a convex glass lens, or the organization of a newspaper advertisement for a pop festival. But the more widespread the events, *C*, become, the rarer such causally correlated events seem to become. Extrapolating on this, it would seem reasonable to suggest that the probability of such causal correlations approach zero as the distance between the events *C* approaches infinity. But this is not to be confused with the statement that the

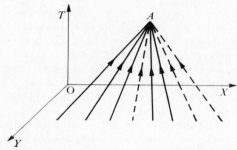

Fig. 26. *An infinite set of processes converge to cause the event* A

probability of the existence of causally *un*correlated events engendering a common effect approaches zero as the distance between the causally *un*correlated events approaches infinity. The first of the two statements seems plausible enough. The second is simply false. The point that Popper, Hill, and Grünbaum would wish to make if I read their intentions correctly is that the *kind* of processes which would have to focus on a galaxy in order to ensure that the galaxy runs backwards, the particular sequences of different *kinds* of radiation needed at all the different points throughout the galaxy, would be extremely rare to the point of having almost zero probability for any particular volume of space.

But what sort of probability is this supposed to be? If the probability is supposed to be epistemic, that is, based on knowledge of our past observations, and no other considerations, then of course this probability will be vanishingly small. For, as explained earlier, we will never observe such occurrences. On the other hand, if the probability is statistical, then it will not be zero but undefined. Why? Firstly, we are dealing with an infinite system with, therefore, an infinite number of possible descriptions of the universe, each with an infinite number of possible sub-microscopic states. Secondly, the sort of events we find to be common in our solar system are probably extremely rare throughout our galaxy anyway. Given a cloud of dust that is going to collapse gravitationally into a star the chances seem infinitesimal that the result would be anything like our solar system, let alone a solar system which would have in it a burning log of wood or a platypus laying an egg. Add together all the sorts of processes that are common within our solar system, and the probability of anything like a repeat performance grows less and less as the detail of the total description grows greater. There does not seem to be any point then, in comparing the probabilities of the occurrence of systems 'like' ours with the probability of the occurrence of systems that are 'like' ours except for being temporally inverted, *unless* we specify precisely the *way* in which these systems are like ours. And even when we specify the way in which the temporally inverted systems are supposed to be like ours, we must take care not to bias our considerations, by using descriptions of sequences of events which are *epistemically* probable within the volume of space and time that we regularly observe and comparing these probabilities with pseudo-statistical 'probabilities' of the inverse description, where these intuited pseudo-statistical 'probabilities' are, more likely than not, the vanishingly low epistemic probabilities of the reverse sequences occurring in this vicinity. And here is another factor about our locality in this universe. Matter is now known to come in two forms. The stuff that we commonly observe around us, is still called 'matter' by physicists. But there is another sort of stuff called 'anti-matter'. The anti-matter is just like matter except that its atomic particles carry reverse charges. Thus, corresponding to the negatively charged electron, there is a positively charged positron. Corresponding to the positively charged proton (which is a matter particle), there is the corresponding anti-matter particle, namely the negatively charged anti-proton. Now when a particle of matter collides with its corresponding particle of anti-matter, both particles vanish and a burst of electromagnetic radiation is generated. The reverse process may also occur. Now the fact is that in this particular locality in space and time there seems to be an over-all preponderance of matter (fortunately for us, otherwise the anti-matter would destroy us). So consider a volume of space and time far greater than that which we have been able to observe, and regard this vast volume as being filled with particles of both matter and anti-matter distributed at random and having random relative velocities, in addition to electromagnetic radiation of all wavelengths moving in random directions. Now consider the chances of a large amount of the matter, as opposed to the anti-matter, becoming, by virtue of the individual interactions of the particles, split off from the rest of the matter, anti-matter, and radiation, to form a relatively closely knit body of cool gas; that is, a gas whose particles have a small enough relative velocity for them to be gravitationally cohesive in the vast sea of empty space that comes by chance to surround them. An unlikely story? Yet it is one explanation for the origin of the system of galaxies that surround us.

The moral of the story is this. We cannot base a belief on the asymmetry of time or even of a pervasive asymmetry of events within time on the implausibility of explanations for the origin of low entropy systems within a larger system of high entropy. That sort of implausibility, if it is at all rational, strikes at the system of galaxies we find ourselves in, just as much as it would any other low entropy system. A system consisting of a cold body absorbing energy being focused in upon it from its hotter surroundings is just as much a low entropy system as a hot body radiating energy to its colder surroundings. The chances of either system evolving from one of high entropy is small given a large but finite spatio-temporal volume. Given an infinite spatio-temporal volume, the chances of either evolving are undefined.

As a conclusion of the last two chapters, then, it would appear that the major difference which we intuitively feel between a temporal dimension on the one hand and spatial dimensions on the other hand, namely that the temporal dimension is inherently directional, is illusory. There are good reasons to believe that it is illusory, there are no good reasons to believe that it is otherwise. The theory of The Present, as a theory which bestows some sort of ontological superiority on some simultaneous sets of events, is rendered doubtful by virtue of the evidence to suggest that simultaneity is relative—not absolute. The theory that later periods of time are in any fundamental way different from earlier periods of time falls by the wayside when we take a wider cosmological view of things, instead of merely extrapolating from the asymmetry of processes that we commonly observe in our immediate vicinity. Practically every theory of physics treats time as being a different sort of dimension from any spatial dimension, but no theory commonly accepted today yields any support for our everyday intuitions, as opposed to the more esoteric considerations of section 4.4, as to where these differences lie.

6.6 SOURCES AND BIBLIOGRAPHY FOR CHAPTER 6

This whole chapter was stimulated to a large extent by the writings of Adolf Grünbaum, especially his *Philosophical Problems of Space and Time* (Routledge & Kegan Paul, London, 1964), Chapter 8, but also by Gale's *The Language of Time* (Routledge & Kegan Paul,

London, 1968), Part III. Reichenbach's causal theory of time is expounded in his *The Philosophy of Space and Time* (Dover, New York, 1957), and Grünbaum's criticism of it is found in Chapter 7 of his *Philosophical Problems of Space and Time*. Reichenbach believed that one could by experiment find out whether one event was earlier than another by examining causes and effects and that one could detect whether or not B was the cause of C, rather than vice versa, independently of knowing whether or not B preceded C. That is, he believed that it was contingently, not analytically, true that causes preceded their effects. Dummett like Reichenbach believes that the matter is a contingent one, and claims that it is logically possible to bring about (that is, cause) some past event. His views on this matter are expressed in his 'Bringing about the past', *Philosophical Review*, vol. 73 (1964). The article is reprinted in Gale's reader *The Philosophy of Time* (Macmillan, London, 1968).

Gettier's article 'Is justified true belief knowledge?', *Analysis*, vol. 23, no. 6 (June 1963), referred to in section 6.3, has not been without criticism, but I think that, in these criticisms, beliefs that are not reliably caused are counted as not being justified. The articles, therefore, in no way impinge upon the causal account of knowledge presented here.

Philosophers have always recognized that knowledge was asymmetrical with respect to time, but they have traditionally regarded this as a contingent matter in need of a contingent explanation. This is clear I think from the writings of Grünbaum in Chapter 9 of his *Philosophical Problems of Space and Time*, where he 'explains' this asymmetry by reference to the fact that we are surrounded by what he calls 'entropic branch-systems'. Gale's Chapter VIII in *The Language of Time* is closer to the mark, I think, but even he seems to think that there is an asymmetry in time to be explained in this area.

For further reading on the Second Law of Thermodynamics, the reader is referred to Allen and Maxwell's *A Text-book of Heat*, Part II (Macmillan, London, 1948). The physicist who popularized the notion that increase in entropy yielded 'time's arrow', as he called it, was Eddington in Chapter IV of his book *The Nature of the Physical World* (Macmillan, New York, 1928). Eddington's idea still survives in popularizations of physics despite numerous criticisms since then. A review of these criticisms may be found in Chapter XLIII of *A Text-book of Heat* and also in Chapter 8 of Grünbaum's *Philosophical Problems of Space and Time*. The notion of entropic 'branch-systems' and their relation to the directionality of time was first introduced by Hans Reichenbach in his *The Direction of Time* (University of California Press, Berkeley, Calif., 1971). Grünbaum's treatment of the subject of branch-systems differs from Reichenbach's in that he does not feel himself obliged to assume, as Reichenbach does, that at any particular time the increase in entropy of all branch-systems throughout the universe is in the same direction. The remarks of section 6.5 were primarily stimulated by part B of Chapter 8 of Grünbaum's book.

Conclusion

THE EXISTENCE OF SPACE AND TIME: THE NATURE OF THE RELATIONALIST PROGRAMME

This book has been concerned with four basic paradoxes regarding space and time.

They are:

1. Necessarily if there is nothing else at a particular place then there is at least space there. So necessarily there is no place where there is nothing. So necessarily there are no vacuums. But of course there *are* vacuums or at least it is a contingent matter whether or not there are vacuums.

2. Necessarily if there are no events occurring or nothing else in existence within a particular temporal duration, then there is at least the temporal duration occurring at that time. So necessarily there are no times when there exists nothing at all. So necessarily there is always something that exists at any time. But whether or not anything exists at any particular time is surely a contingent matter.

3. Necessarily space and time are passive. Events take place within space and time, but space and time themselves do not partake causally in the events which occur within them. But if that is so, our beliefs about space and time are causally unrelated to the existence of space and time. So we don't *know* that space and time exist. But of course we do know that space and time exist.

4. Necessarily time flows. But if anything flows, then it flows with respect to time. So time flows with respect to itself. But nothing can flow with respect to itself.

The relationalist programme was defined as an attempt to rid us of these paradoxes by reducing statements which seemed truly to ascribe certain properties to space or time to statements which made no such ascription—thereby ridding us of the need to refer to space and time. Throughout this book so far, I have been following a relationalist programme. Sometimes appropriate reductions have been achieved, sometimes not. I now wish to show that the relationalist programme can always be made to work. The exact details may escape us in a particular case, but there is always a possible reduction of expressions involving 'space' and 'time'. The proof which follows is for the case of space alone, but a similar proof with the word 'space' replaced by 'time' throughout, would be applicable to time.

Let us assume that no reduction has been achieved for properties Φ_1, Φ_2, . . . , Φ_n of space, that is, the statement:

(1) All space is Φ_1 and Φ_2 and . . . and Φ_n

is not reducible in any of the senses of reduction which were mentioned in section 1.4.

Let us assume, furthermore, that there are no other irreducible properties of space. Then it will be true also that if anything is Φ_1 and Φ_2 and . . . and Φ_n it will be space.

For convenience let us write 'Ψ' for the complex description 'Φ_1 and Φ_2 and . . . and Φ_n'. Then,

(2) Space is Ψ and only space is Ψ.

will be true.

Now (2) will either be analytically true or contingently true. If it is analytically true, then an analytic reduction of all expressions using 'space' is possible by replacing 'space' by 'that which is Ψ'. (2) itself will reduce to the tautology:

(3) That which is Ψ is Ψ and only that which is Ψ is Ψ.

That such a reduction is possible, is, of course, contrary to our assumption that space being Ψ was irreducible. So (2) must be contingently true, that is, its truth is dependent on the way the world happens to be. Let us assume, then, that (2) is contingent.

Now space being Ψ is either causally efficacious with respect to some events involving matter or it is not. If it is not, then there is no way in which we can find out that space is Ψ, so we will not know that space is Ψ and we will have no reason to believe that there is something, namely space, which is Ψ. In which case an ontological reduction of space is as much in order as is an ontological reduction of phlogiston. Now let us consider the case when space being Ψ *is* causally efficacious. Ψ may or may not be a property which varies either in strength or direction or both from one place to another and from time to time. Thus a person's scalp may everywhere be covered in hair, but in some places the density of hair will be greater than others. But let us assume firstly, that Ψ is not like this, that is, we shall assume that if it is a quantitative and/or directional property, then its magnitude and direction are universal constants. This is the case considered in section 3.4, but I shall repeat the gist of the argument here for the sake of completeness.

If space *were* ubiquitously Ψ then we could never set up an experiment in which space was not Ψ. That is, any evidence we would ever have that distributions of matter of type E, together with space being Ψ, always cause distributions of matter of type E^1, would also be evidence for the simpler statement:

Distributions of matter of type E always cause distributions of matter of type E^1.

Thus in this case, space being Ψ might well be causally efficacious, but we would have no reason to suspect that that was so. But if we had no reason to believe that space being Ψ was causally efficacious, then we would have no reason to believe that space being Ψ caused anything including any beliefs that people held to the effect that space was Ψ. That is, we would have no reason to believe that any-one *knew* that space was Ψ. In which case, an ontological reduction would once again be in order. However, as was illustrated in section 3.4, the relationalist's reduction may well be rejected in order to retain the unifying explanatory power of the theory that space is Ψ, but alternative unifying explanatory theories could, in principle, if not in practice, be offered to replace the theory that space is Ψ.

Now let us assume that Ψ *is* a quantity which varies in value and/or direction from place to place. Let the various values and/or directions that Ψ can have at various times and places be Ψ_1, Ψ_2, . . ., Ψ_n. Then if the value of Ψ that space possesses at a particular place and time is causally efficacious, it will not in general be true that any evidence we have that distributions of matter of type E, *together with space being* Ψ_1, say, will always cause distributions of matter of type E^1, would also be evidence for the simpler statement that distributions of matter of type E always cause distributions of matter of type E^1. A case of a type E distribution which is also a case of space being Ψ_1 may yield a different effect from a case of type E distribution which is also a case of space being Ψ_2. Experiments wherein the type of distribution of matter was kept constant while the value of Ψ was varied would be possible, and such experiments would yield evidence for the causal relevance of the variable quantity Ψ of space.

But, in reply to this, the relationalist could always say that space is, contrary to the above considerations, causally inefficacious, or as he would prefer to put it, that causal laws are spatially and tem-porally uniform, and that what the above considerations show, therefore, is that some material pervades the whole of space, and it is this material (aether or field) which bears the property of having the quantity Ψ which varies from one part of this all-pervading substance to another. In such a case the relationalist would have to agree with Melissus that there were no vacuums, but he would not feel himself obliged to believe that this was necessarily the case.

The same considerations apply *mutatis mutandis* throughout the discussion for time, and the relationalist might, given certain experimental evidence, be forced to conclude that there was always some matter in existence, but he would not have to conclude that this was necessarily the case.

Again with respect to the epistemic paradox (Number 3) he could simply reduce the initial premiss, namely 'Necessarily space and time are passive', to 'Necessarily causal laws are spatially and temporally uniform'. The remaining premiss and the conclusion are then unrelated to this replacement and the argument falls apart. He could agree that we know that space and time exist, but would suggest that this is merely a misleading way of expressing our knowledge with respect to some statement which does not have to be expressed by seeming to refer to some entities called 'space' and 'time'.

Of course an absolutist might jib at any one of the ontological reductions which the relationalist is willing to introduce but, because they are ontological reductions rather than analytic reductions, the issue will be more than a mere verbal dispute in these cases. Often the dispute will involve a clash between experimental evidence and one or more fundamental principles of physics. Thus let us reconsider for a moment the case where the absolutist wished to claim, on the basis of reasonable experimental evidence, that space had a property Ψ which varied in magnitude and direction from place to place and time to time. The relationalist saw this as being contrary to the principle that causal laws relating distributions of matter were spatially and temporally uniform. The absolutist preferred to accept this rather than to reject, say, his atomism, that is, rather than accept the idea of an all-pervading material plenum. Both these positions are consistent with and indeed imply the principle that matter, time, and space are quite distinct. In his development of the General Theory of Relativity Einstein suggested that this is the principle which should be thrown overboard.

It is clear that because evidence is involved, and because ontological reductions are being offered, the issue is not simply verbal. But that does not mean that the issues are decidable by a few isolated experiments—whatever their results. Refurbishments of our conceptual schemes which allow for greater diversity in the hypotheses we are able to consider may be essential. The first requirement for a modern theoretical physicist is that he or she be broad-minded.

Selected Bibliography

(A more extensive bibliography on space and time may be found on pages 427–36 of J. J. C. Smart's reader *Problems of Space and Time*, which is listed below.)

Aleksandrov, A. D., Kolmogarov, A. N. and Lavrentev, M. A., *Mathematics, its content, methods and meaning*, vol. 3. Translated by Hirsch, K. A., M.I.T. Press, Cambridge, Mass., 1969.

Alexander, H. G., *The Leibniz–Clarke Correspondence*, Manchester University Press, Manchester, 1956.

Allen, H. S. and Maxwell, R. S., *A Text-book of Heat*, Part II. Macmillan, London, 1948 (especially Chapters XXIX, XXX and XLIII).

Bridgman, P. W., *The Logic of Modern Physics*, Macmillan, New York, 1927.

Broad, C. D., *An Examination of McTaggart's Philosophy*, vol. 2, Part 1. Cambridge University Press, Cambridge, 1938 (especially Chapter 35).

Descartes, R., *Principles of Philosophy Part II*. English translation in Anscombe, E. and Geach, P. T. (eds.), *Descartes: Philosophical Writings*, Thomas Nelson & Sons, Edinburgh, 1954.

Earman, J., 'Who's afraid of Absolute Space?', *Australasian Journal of Philosophy*, vol. 48, no. 3 (December 1970).

Eddington, A. S., *The Nature of the Physical World*, Macmillan, New York, 1928 (especially Chapter IV).

Einstein, A., *Relativity—The Special and the General Theory*, 15th edition, translated by Lawson, R. W., Methuen, London, 1954.

Einstein, A., 'Autobiographical Notes', in Schilpp, P. A. (ed.), *Albert Einstein Philosopher—Scientist*, vol. 1. Harper & Row, New York (Harper Torchbook edition, 1959).

Ellis, B., *Basic Concepts of Measurement*, Cambridge University Press, Cambridge, 1968.

Frank, N. H., *Introduction to Electricity and Optics*, 2nd edition, McGraw-Hill, New York, 1950.

Gale, R. M., *The Language of Time*, Routledge & Kegan Paul, London, 1968.

Gale, R. M. (ed.), *The Philosophy of Time*, Macmillan, London, 1968.

Goodman, N., *The Structure of Appearance*, Harvard University Press, Cambridge, Massachusetts, 1951.

Grünbaum, A., *Philosophical Problems of Space and Time*, Routledge & Kegan Paul, London, 1964.

Mach, E., *The Science of Mechanics: A Critical and Historical Account of its Development*, translated by McCormack, T. J., Open Court, La Salle, Illinois, 1960 (especially Chapter II, Section 6).

McTaggart, *The Nature of Existence*, Book V, Chapter 33, Cambridge University Press, Cambridge, 1927.

Moore, G. E., *The Commonplace Book of G. E. Moore* (ed. Lewy, C.), Allen & Unwin, London, 1962. Notebook II (C 1926), entry 8.

Nagel, E., *The Structure of Science*, Routledge & Kegan Paul, London, 1961 (especially Chapters 8 and 9).

Newton, I., *Mathematical Principles of Natural Philosophy*, Florian Cajori (translator), University of California Press, Berkeley, Calif., 1934.

Panofsky, W. K. H. and Phillips, M., *Classical Electricity and Magnetism*, Addison-Wesley, Reading, Mass., 1955.

Poincaré, H., *The Foundations of Science*, translated by Halstead, G. B., Random House, New York, 1970.

Popper, K. R., 'Philosophy of Science: a Personal Report' in Mace, C. A. (ed.), *British Philosophy in Mid-Century*, Cambridge University Press, Cambridge, 1957.

Prior, A. N., *Past, Present and Future*, Clarendon Press, Oxford, 1967.

Quine, W. V. O., *Word and Object*, Wiley, New York, 1960.

Reichenbach, H., *The Direction of Time*, University of California Press, Berkeley, Calif., 1971.

Reichenbach, H., *The Philosophy of Space and Time*, Dover, New York, 1957.

Reichenbach, H., *Elements of Symbolic Logic*, Macmillan, New York, 1947 (especially Section 50).

Smart, J. J. C. (ed.), *Problems of Space and Time*, Macmillan, New York, 1964. (The Bibliographical notes in this reader are very extensive.)

Smart, J. J. C., *Between Science and Philosophy*, Random House, New York, 1968 (especially Chapters 7 and 8).

Smart, J. J. C., *Philosophy and Scientific Realism*, Routledge & Kegan Paul, London, 1963 (especially Chapter VII).

Van Fraassen, B. C., *An Introduction to the Philosophy of Time and Space*, Random House, New York, 1970.

Index